CLIMATE VARIATIONS, DROUGHT AND DESERTIFICATION

by

F. Kenneth Hare

revised by

L. A. J. Ogallo

WMO – No. 653

WORLD METEOROLOGICAL ORGANIZATION

1993

Cover: Changes in vegetation index in Africa from satellite images prepared by FAO-RSC ARTEMIS. False colours show greenness of surface. Reds and purples indicate wettest conditions, tans and browns the driest.

WMO – No. 653

ISBN 92-63-12653-4

NOTE

The designations employed and the presentation of material in this publication do not imply the expression of any opinion whatsoever on the part of the Secretariat of the World Meteorological Organization concerning the legal status of any country, territory, city or area, or of its authorities, or concerning the delimitation of its frontiers or boundaries.

Contents

		Page
Foreword		v
Chapter 1:	Introduction	1
Chapter 2:	A measure of aridity: the dryness ratio	5
Chapter 3:	What is desertification?	8
Chapter 4:	Recent climatic fluctuations	13
Chapter 5:	Causes of climatic variation	17
Chapter 6:	Climatic feedbacks: Does human interference worsen drought?	20
Chapter 7:	Climate and the control of desertification	24
Chapter 8:	Climate and the future	28
Annex I :	Strategy for WMO activities in support of the combat against drought and desertification during the post-UNCED period	35
Annex II :	Guidelines for strategy regarding the implementation by WMO Members of UNCED Agenda 21, Chapter 12 on desertification and drought	38
Abbreviations		45

Foreword

Following the United Nations Conference on Desertification (UNCOD) (Nairobi, August September 1977), the UN General Assembly adopted Resolution 32/172 of December 1977 and a Plan of Action to combat desertification. Recommendation 23, in particular, invited the World Meteorological Organization (WMO) to support the international actions to combat desertification in the context of the UN Plan of Action. This recommendation was considered by WMO's Executive Council at its thirtieth session (May 1978), which approved a WMO Plan of Action that included activities aimed at publicizing the meteorological and hydrological aspects of droughts and desertification.

In the framework of these activities, a booklet entitled "Climate variations, drought and desertification" was prepared by Prof. F. Kenneth Hare (Canada) in connection with the celebration of World Meteorological Day on 23 March 1985. World Meteorological Day is celebrated each year to commemorate the coming into force on 23 March 1950 of the WMO Convention and the transformation of the former non-governmental International Meteorological Organization into the intergovernmental World Meteorological Organization, as a specialized agency of the United Nations.

The booklet focused on the theme chosen for the celebration of World Meteorological Day 1985, highlighting the importance of climate, climate variations and drought, and WMO activities in the combat against the effects of drought and desertification. It also described WMO efforts in the implementation of the UN Plan of Action.

The UN Conference on Environment and Development (UNCED) that was held in Rio de Janeiro in 1992 called for the establishment of an Intergovernmental Negotiating Committee for the elaboration of an international convention to combat desertification in those countries experiencing serious drought and/or desertification, particularly in Africa, and requested that this convention should be finalized by June 1994. It was therefore considered appropriate at this time to revise and update the booklet prepared by Prof. Hare, and to make available the updated version as a suitable aid to the negotiation process. Others not involved in the negotiation but interested in the subject of climate variation and drought in desertification processes will find this updated edition useful.

This important task of revising and updating the booklet was undertaken by Prof. L.A.J. Ogallo (Kenya), to whom I wish to express my deep gratitude. It is my hope that the contents of this booklet, which should be considered as one of the WMO contributions to the negotiating of the convention, will enable the scientific community as well as national authorities to reflect on the issues of climate variations as they relate to drought and desertification. It may also help nations, particularly those experiencing drought and/or desertification, to pursue the negotiations on the elements of the desertification convention with a clearer understanding of the interactions of climate and desertification.

G. O. P. Obasi
Secretary-General

CHAPTER 1

Introduction

Deserts are regions of the world with too little rainfall to support most forms of life. Around these deserts and over many sub-tropical parts of the world are found semi-arid and sub-humid lands which support a large human population and unique species of plants and animals. Many national States lie within these areas of deficient rainfall. In spite of all their hazards, these drylands — deserts and surrounds alike, with their wild climates — have offered many challenges to humanity, many of which have for several centuries been successfully answered using the lessons and survival techniques of the previous generations of their ancestors. Several of the dominant cultures and religions of the modern world — Islam, Judaism and Christianity, for example — had their origins in the arid zone.

Yet today these drylands are in many places the scene of much acute distress and even of tragic famine. In spite of the long traditions of human adaptation, many nations in these areas now find themselves unable to wrest a reasonable living from the vanishing soils and natural resources. Although climate-induced stress has been common in nature, occurring on all the continents, including North America, Australia, Europe and Asia, the highest degree of distress and vulnerability has been witnessed in the developing parts of the world, especially in Africa, where many climate-related disasters and much misery have been reported in recent years. Some areas have now become more prone to famine, with consequent large-scale abandonment of previously fertile land. Much damage has been inflicted on the national economies, leading to a lot of hardship for the majority of the people in the drylands. Very few parts of the arid zone have been spared.

What accounts for this unhappy situation? Is it human failure, with climate being used as a scapegoat? Are the causes to be found in human interference with the ecosystem? Or is it due to a normal change in the climate, induced by natural processes in the climate system?

In part the answers to these questions involve the problem of desertification, the spreading of desert-like conditions into formerly productive land or, more correctly, the degradation of that land until it can no longer adequately support living communities. So widespread is this degradation that in 1977 the United Nations organized in Nairobi a Conference on Desertification (UNCOD). The conference examined the problem, found it to be a reality on every continent, and adopted a number of recommendations. These recommendations were accepted by the UN General Assembly, which adopted Resolution 32/172 of December 1977 and a Plan of Action to combat desertification.

The complex impact of land degradation and desertification processes on the climate has been recognized by the WMO/UNEP Intergovernmental

Panel on Climate Change (IPCC) in a report which provided the background information for the Framework Convention on Climate Change, which has been signed by many countries. Apart from the climate change convention, four other important relevant decisions came out of the recent United Nations Conference on Environment and Development (UNCED), which was held in Rio de Janeiro, Brazil, in June 1992. These are the Rio Declaration, the Biodiversity Convention, the Forestry Principle Statement and Agenda 21. In addition, UNCED recognized drought and desertification as important issues which must be addressed by the UN General Assembly since they are the cause of much human suffering and hamper development in the drylands. Chapter 12 of Agenda 21, especially paragraph 12.40, recommended to the UN General Assembly at its forty-seventh session the establishment of an Intergovernmental Negotiating Committee (INC), under the aegis of the General Assembly, to elaborate an international convention to combat desertification in those countries experiencing serious drought and/or desertification, especially in Africa, with a view to finalizing such a convention by June 1994. Agenda 21 highlights preventive measures for halting the spread of the deserts over land areas which have not yet been degraded or which are only slightly degraded. It also described various ways of dealing with land areas which have been severely affected by desertification.

The World Meteorological Organization has played a significant role in the past, jointly with other UN bodies, and has subsequently contributed to the efforts to control desertification.

Desertification affects an estimated 70 per cent of all of the drylands of the world, amounting to about 3.5 billion hectares. About one-sixth of the world population, living in these areas, is affected by desertification processes.

Why do the deserts seem to be spreading? The answer is twofold.

Firstly, the human pressure in the dry zones has grown enormously in recent decades due to an increase in population. The need for food, water, fuel, raw materials and other natural resources has grown accordingly. In many dryland areas the demands on the soil, vegetation and climate now greatly outweigh their capacity to yield: the carrying capacity of the land has been exceeded.

Secondly, many recent years have seen protracted drought, sometimes lasting for over twenty years. Under natural conditions such failures of expected rainfall would have had little effect. Human society has been adept at overcoming drought by using indigenous adaptation methods inherited from generation to generation. But the rise in population and disappearance of various forms of regulation have made this adaptation more difficult. Nomadic pastoralism, for example, has declined as a livelihood system, together with the practice of shifting cultivation. Drought has, in some areas, resulted in the loss of production, property and life, in famine, diversion of national development programmes and many socio-economic disasters. The mid-1980s drought in sub-Saharan Africa, for example, led to the death of about three million people.

Droughts have in some areas driven nomadic pastoralists into the zones of cultivation. New population settlements with new but drought-vulnerable land-use activities have also increased in drought-prone areas on account of population pressure. These activities have resulted in various conflicts between traditional land-use activities.

The term *arid zone* has been used in this text as a collective expression for the world's drylands, including deserts, savanna, dry forests and semi-desert scrubs. It comprehends at least 30 million square kilometres of the land surface of the globe and by some definitions over 50 million. The arid zone has been a stage centre of many UN activities, including those of WMO, for many years now.

This booklet explores the climatic stresses that underlie the tensions just described. Drought is obviously common in the dry climates; in a sense, the latter are areas of perpetual drought. Properly defined, however, climatic drought is the failure of expected rainfall; it is hence a shock to any economy that is already overstressing its food producing system and natural resources.

Certain broad themes will run through this entire publication and it will be of help to the reader to keep it in mind that:

- Desertification is to be seen in the light of overall ecosystem change, including the effect of variable climate;
- Climate anomalies have recently subjected many systems, especially those in the arid zones, to very severe stresses;
- Climate is an interactive system, in which atmosphere, the Earth's living cover and human enterprise affect one another;
- In particular, certain kinds of land and resource use may affect climate by means of feedback processes;
- Climate changes can also affect the quality and quantity of ecosystems, and all kinds of land- and resource-use activities;
- Sustainable land-use and water management policies, environmentally sound agricultural and pastoral techniques, accelerated afforestation and reforestation programmes using drought-resistant and fast-growing plant species, indigenous knowledge of the forests, rangeland, forest lands, and natural vegetation, and good knowledge of the interactions between natural and human forces are crucial in any plans to combat desertification;
- Climatologists have a role to play in combating desertification; this role should be recognized by national governments and be internationally guided;
- Future climate may well be different from that of today and it is vital to foresee just how different;
- Political will plays a crucial role in the development and effective implementation of any technical programme.

The object is to present a balanced perspective regarding the role of climate in the uses made of the arid zone by human society. The major finding will be that climate alone will not destabilize the productive cloak of life that shelters the world and yields so much to human needs — but, nevertheless, it cannot be disregarded.

CHAPTER 2

A measure of aridity: the dryness ratio (aridity index)

In discussing the climate of the arid world one needs some comparative measures of aridity. Only two of the most common measures of aridity will be highlighted in this text, namely the Budyko-Lettau dryness and P/PET ratios. The Budyko-Lettau ratio expresses the ratio of net radiant energy at the Earth's surface to the heat required to evaporate a year's rainfall. In plain language, a value 1 for this index indicates that net radiative heating of the surface is sufficient to evaporate one year's rainfall. This value is observed close to the limit between forest and grassland.

In a comparative review the UN Conference on Desertification found that the zone most at risk to desertification processes lies between the dryness ratios of 2 and 7. At higher values the biological productivity is already too low to support the large human populations which cluster around or along the main oases. At values below 2 the remaining vegetation is generally vigorous enough to recolonize the damaged land. In the moister regions, however, soil erosion is a serious problem.

The P/PET ratio expresses precipitation (P) as a fraction of the Potential Evapotranspiration (PET). PET represents the maximum amount of moisture which could be evaporated at the surface and transpired by vegetation, if it were available.

Table 1 gives some classification guidelines used by UNEP to map the dry climates of the world presented in the centerfold, map C.

Table 1—**Aridity zones by P/PET ratios** *(UNEP, 1992)*

Climate zone	R = P/PET ratio	% of world covered
Hyper-arid	$R < 0.05$ Very limited human activity around oases	7.5
Arid	$0.05 \leq R < 0.20$ Pastoralism is possible but highly susceptible to inter-annual climate variability or groundwater resources	12.5
Semi-arid	$0.20 \leq R < 0.50$ Sustainable pastoralism. Agriculture susceptible to high degree of interannual climate variability	17.5
Dry sub-humid	$0.50 \leq R < 0.65$ Wide range rainfed agricultural activities practised	9.9
Moister climates	$R \geq 0.65$	
– Humid		39.2
– Cold		13.6

The deserts of the world

Deserts are regions of the world where there is too little rain for most forms of life. Without adequate rain, plants cannot grow properly and hence do not provide enough food for animals. Large parts of the central and northern Sahara are virtually rainless, as is much of the Atacama Desert of Chile as well as a few regions on other continents. These rainless areas are true deserts, occupied by dry rock surfaces or expanses of shifting sand. Life is banished to places where groundwater seeps to the surface — i.e to oases — or to hilly areas perpetually bathed in mist.

More extensive than the deserts themselves are the wide areas of semi-desert scrub, savannahs and dry forests that surround them. These semi-arid or sub-humid areas normally experience a short rainy season, marked by a few torrential rainstorms every year, during and after which the local ecosystems work furiously to achieve their year's quota of new tissues (biomass). For a few weeks to a few months the biological productivity is high. Green plants use the newly arrived moisture to grow rapidly, and animal populations eat their fill.

A few weeks after the end of the rains, the verdancy and activity disappear. The land becomes dry and dusty. Fire is common. Animals have to adopt elaborate strategies to survive. The system awaits the next annual round of rainfall, subdued in productivity, but well able to spring back into furious life when water reappears in the soil. Dry ecosystems are hence secure and stable, and are usually able to outwait even the total failure of a rainy season. Anomalies from one year to another are however sometimes very large.

Where are these formations to be found?

On all five continents, but unequally distributed (see centerfold, map C). The arid zone is mainly subtropical, because it is caused by atmospheric subsidence (downward motion of the air) which suppresses rainfall. Most of these areas are close to latitudes 30° north and south. The major areas are:

(*a*) In Africa, the Sahara Desert, which has scrub forests along its northern limit. To the south is a belt of semi-desert scrub and sparse savannahs (the Sahel) and a larger extent of savannahs and dry forests including much of East Africa, Angola, Botswana (the Kalahari), parts of the Republic of South Africa and south-western Madagascar. Many parts of this vast area are severely affected by desertification;

(*b*) A wide belt of temperate steppes, semi-desert and desert extending from the Middle East across Uzbekistan, Kazakhstan, Kyrgyzstan, Tajikstan and Turkmenistan to China, and including north-western India, Pakistan, much of Iran and Afghanistan. Many developed nations have undertaken elaborate measures against desertification. Pakistan has been deeply concerned with the salinization of her irrigated territories while India has a research institute in the Thar desert area devoted to arid zone studies;

(*c*) In South America, the Atacama Desert of Chile, together with extensive savannahs and dry forests in Argentina, Paraguay and Bolivia. There is also an area of savannahs and dry forests in north-eastern Brazil, an area of acute population pressure and frequent drought;

(*d*) The south-western United States and northern Mexico, where near-desert conditions occur in parts of Sonora, New Mexico, California and Nevada. A broad area of semi-desert scrub, dry forest and steppe (dry grassland) surrounds the belt;

(*e*) Much of the interior of Australia, which has dry forests, eucalyptus and acacia savannahs and some poor native grassland, with extensive areas of shifting desert sands. Desertification is widespread.

Such, then, is the arid zone with its pervasive problem of progressive decay of natural ecosystems under the impact of increasing populations and hostile episodes of climate. It is clearly incorrect in most cases to speak of the "remorseless advance of the desert". This does indeed happen in some areas — for example in western China, Australia and the Sudan — because of advancing sand dunes; but most desertification is the on-the-spot degradation of productive ecosystems by such agents as pounding hooves, traffic and even human recreation (as in California and Arizona).

The hostile forms of human land use are many and varied. The substitution of domestic animals for the migrating herds of big game so typical of the African savannahs, for example, means that drinking water must be provided for stock (since domesticated species lack the capacity to derive enough water from plant tissues, or to store it for long periods). This guarantees a stampede towards wells when drought strikes — with resultant destruction of soils and vegetation around them. Human economies, especially those derived from elsewhere, as are many forms of cash-crop agriculture, are often ill-adapted to the realities of the arid zone. This is especially true during episodes of severe drought.

CHAPTER 3

What is desertification?

The development of desert-like conditions where none existed previously has been described in many ways. UNCED defined desertification as land degradation in the arid, semi-arid and sub-humid areas resulting from various factors, including climatic variations and human activities. Desertification is a widespread but discrete process of land degradation in space throughout the drylands. It takes many forms, but usually involves:

- *A reduction in the fraction of the soil covered by vegetation.* Arid and semi-arid ecosystems usually expose some extent of bare soil in the dry season. With desertification, the fraction of bare soil increases and vegetation may be reduced to isolated patches or clumps;
- *A consequent rise in the reflective capacity (albedo) of the surface* for solar radiation, since arid and semi-arid soils are lighter-coloured than most plants, even with the grey foliage so usual in these climates. Hence bare soils are better reflectors. Deserts show up brightly in satellite pictures because of reflected sunlight;
- *A considerable and often permanent loss of perennial plants*, especially woody shrubs and trees;
- *Considerable soil erosion and impoverishment*, because of removal by wind of fine mineral and organic material and because of rapid oxidation of the remaining litter and soil carbon. Gully and sheet erosion of soils by occasional heavy rainfalls tends to accumulate the eroded material on valley floors or in basins.

Causes of desertification

It is obvious from the previous discussion that desertification is caused by two major factors—namely, natural forces, through periodic stresses of extreme and persistent climatic events such as drought; and human use and abuse of sensitive and vulnerable dryland ecosystems. The feedbacks between the natural and human-induced causes, however, are neither well known nor understood.

Climatic anomalies, with significant changes in the temporal and spatial distribution, may result in the lengthening of aridity phases, higher temperatures and winds of greater intensity. Similarly, increasing human pressure may result in over-use of the dryland natural resources and in extension of the cultivated area beyond the limits where man-environment equilibrium can be properly maintained. Such human pressure normally includes the extension of irrigated areas and improper land-use practices, resulting in land degradation through water and wind erosion, overgrazing by livestock, deforestation (for firewood and building), bush and forest fires, alkalinization, waterlogging, etc.

Almost as destructive, but spatially less extensive, is the process of salinization, whereby mineral salts become concentrated in the surface soil by the evaporation of saline soil solutions or through the use of saline water for irrigation. In humid climates such salts are flushed through the soil by excess rainfall, which percolates to the subsoil or escapes to rivers. Such flushing mechanisms are inexistent in the arid zone. Salinization is especially significant in irrigated lands where water management is imperfect and in regions of rainfed agriculture where formerly there was natural grassland.

None of the above processes alone seriously affects natural ecosystems, which are able to cope with them. The entire structure of these systems is adapted to the scarcity and seasonal concentration of rainfall. Damage occurs when human misuse of land combines with the occurrence of drought. The natural mechanisms of repair and renewal cannot then cope with the added stresses. Map D of the centerfold gives the desertification map of the world.

The details of the desertification processes differ from country to country or region to region. The distribution of soil degradation processes over Africa is shown in Table 2 and Figure 1 overleaf.

The distribution and density of human population is one of the major factors which has greatly influenced human-induced land degradation processes in Africa.

Table 2 — **African susceptible drylands by percentage and degree of soil degradation** *(UNEP, 1992)*

Degree of land degradation	% of Africa
Extreme	0.2
Strong	5.7
Moderate	8.7
Light	11.2
Non-degraded	74.2

Drought and the causes

Many definitions have also been given for the term drought. Most of them express drought as a temporary situation where the demand for water exceeds the water supply available from the various resources. Precipitation is the major component in the water supply-demand equation. *Meteorological drought* is therefore said to occur when there is a prolonged absence or marked deficiency or poor distribution of precipitation. This can

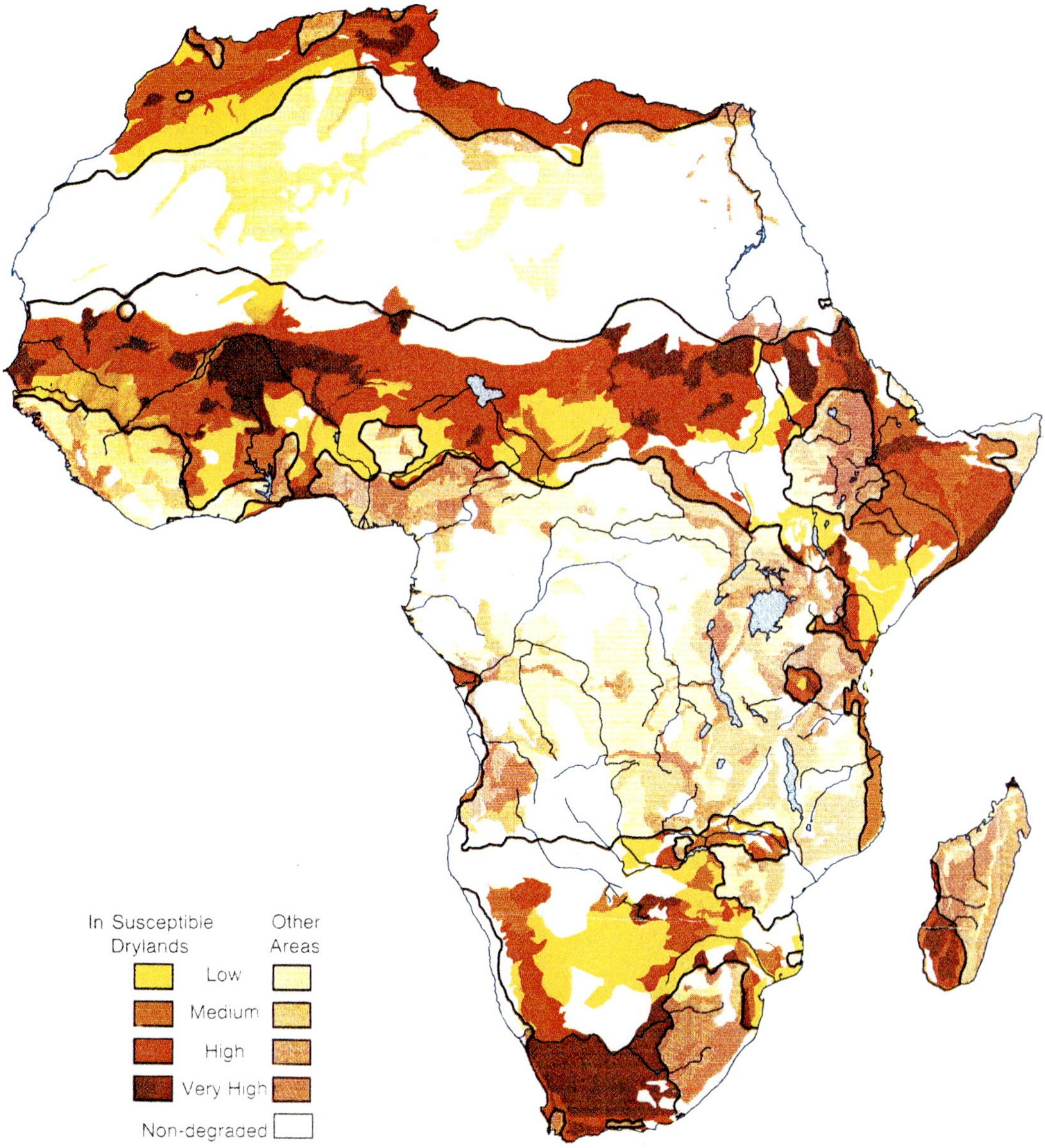

Figure 1 — Soil degradation in Africa (*Source : UNEP/ISRIC CRU/UEA)*

induce severe impacts on many water-use activities. Many of the other definitions of drought have therefore attempted to quantify the impacts of the water deficiency induced in the individual water-use systems by meteorological drought. Examples are *agricultural drought* and *hydrological drought*.

Drought is therefore a temporary feature caused by anomalies in the usual climate of the region. It occurs in virtually all climatic regimes, but with a higher frequency and probability in the semi-arid and sub-humid regions. Meteorological drought should therefore be seen as a major cause of desertification, especially if it persists for several consecutive years, leading to a drastic adverse impact on human activity.

The natural ecosystems of the dry areas of the world have adapted to the wild climate of these zones by using special survival strategies derived from preceding generations. Such strategies have been used to overcome drought hazards for several centuries. The following are especially prone to human disturbances:

(*a*) *Semi-desert scrub* tends to have several distinct components. There is a layer of shrubs, usually scattered in clumps or growing as isolated individuals, often protected against drought by specialized anatomy, such as water-retentive leaves and stems and deep, water-seeking roots. The shrubs may also be thorny or bitter — tending to resist browsing animals. The cactus and mesquite stands of Arizona and the thorn bushes of the African savannahs exemplify these adaptations. Within the shrub stands there is a discontinuous layer of perennial herbs and grasses. Very short-lived ephemeral annual plants, capable of completing their life cycle in perhaps 12 weeks after rain, may be abundant on the otherwise bare soil patches. The latter are often protected against wind and water erosion by a surface crust a few centimetres thick. Such formations are very vulnerable to overgrazing or browsing. The areas of bare soil are then extended, and the protective crust is destroyed by hooves; soil porosity is also reduced. The use of woody shrubs for fuel, or for dry-season animal fodder, further aggravates the vulnerability.

(*b*) *Savannahs* are grasslands across which shrubs and trees are scattered, the latter especially along seasonal watercourses. The grasses are typically tall, fast-growing species that flower, seed and become dormant during and immediately after the wet season. For the remainder of the year they are brown and tinder-dry, when they burn regularly and easily. They also lose much of their nutritional value when dry. Herdsmen may supplement their animals' diet by cutting green tree branches to the point where the trees die. Over much of the arid zone (e.g. in South America, Africa and Australia) acacia trees provide shelter and useful commercial products (such as gum arabic in West Africa). Savannahs occur in the climatic zone where the wet season is long enough to allow growth for two to six months. A characteristic feature is that little biomass carries over from one wet season to the next, except in tree patches, because the grasses die back in the dry season to a small living rootstock. The major areas of desertification were originally mostly savannahs or dry forests in their natural state.

(*c*) *Dry forests* were formerly widespread in the arid zone, chiefly on hilly ground and along the margins of the tropical rainforests. There were comparable woodlands along the poleward margin of many tropical deserts, notably in the Mediterranean countries, California and parts of Australia. Some dry forests are made up of trees that shed their leaves in the dry season (as in parts of Africa and north-western India). Others are evergreen, e.g. the eucalyptus-acacia forests of Australia. All are very prone to fires, but many tree species are resistant to ground fires, which are almost annual events. Most of these once extensive forests have been cleared for agriculture, exploited for firewood or for building, or overpastured; few can be found in an undisturbed state.

CHAPTER 4

Recent climatic fluctuations

Anomalies in the usual climates of individual areas have been observed in many parts of the world. Some of these anomalies covered only small areas and lasted for a short duration. Others have been very severe in magnitude, extending over many regions and persisting for a long period of time. The best known is the so-called Sahelian drought, which has been a period of progressive rainfall desiccation that began more than two decades ago. Periodic occurrences of near-normal rainfall were however recorded over parts of the Sahelian zone within this period of declining rainfall trend. Severe droughts have also afflicted north-eastern Brazil, western China, eastern Australia and many other parts of the world. Many of these episodes have caused great human suffering, especially in Africa.

Figure 2 shows the recent course of annual rainfall across some parts of the world.

It can be seen at once that rainfall varies greatly from year to year in all areas. At any individual dry location, the variability is generally higher. This erratic behaviour of the rainfall is one of the reasons behind the widespread but diminishing use of nomadic pastoralism as a livelihood system in the dry climates. Herders can move their animals north-south with the seasons, but using different strategies from year to year. The variable rainfall also means that cultivators near the dry margin will have to be equally flexible: they must plant or seed the land at the right moment, according to the dates when the rains permit it, and must also be able to absorb the shock of frequent crop failure, should the rains fail. All these climates cry out for versatility, flexibility and the maintenance of reserves against hard times — characteristics of many traditional systems, not only in Africa, but throughout the arid zone. Herein lies the need for application of agrometeorological information in the planning and operations of agriculture in these areas.

The spectacular feature of Figure 2, however, is the prolonged dry spell in recent years over sub-Saharan Africa. The figure indicates that rainfall was generally adequate or abundant in the Sahel region during the 1950s and 1960s, at least until the mid-decade. Then there began a desiccation — brought about by a decrease in rainfall — that led to intense drought in 1972–1973, and to continuing drought in the late 1970s and early 1980s. Near-normal conditions appeared in the late 1980s. The result of extended drought has been the depletion or exhaustion of soil and shallow groundwater, and intense pressure on farmers, herders and all those with associated occupations. The Sahelian drought, once perceived as a severe but short event (1968–1973) in line with many previous drought episodes, has turned out to be a prolonged desiccation that has brought tragic famine and dislocation to the people and immense damage to Africa's ecosystems.

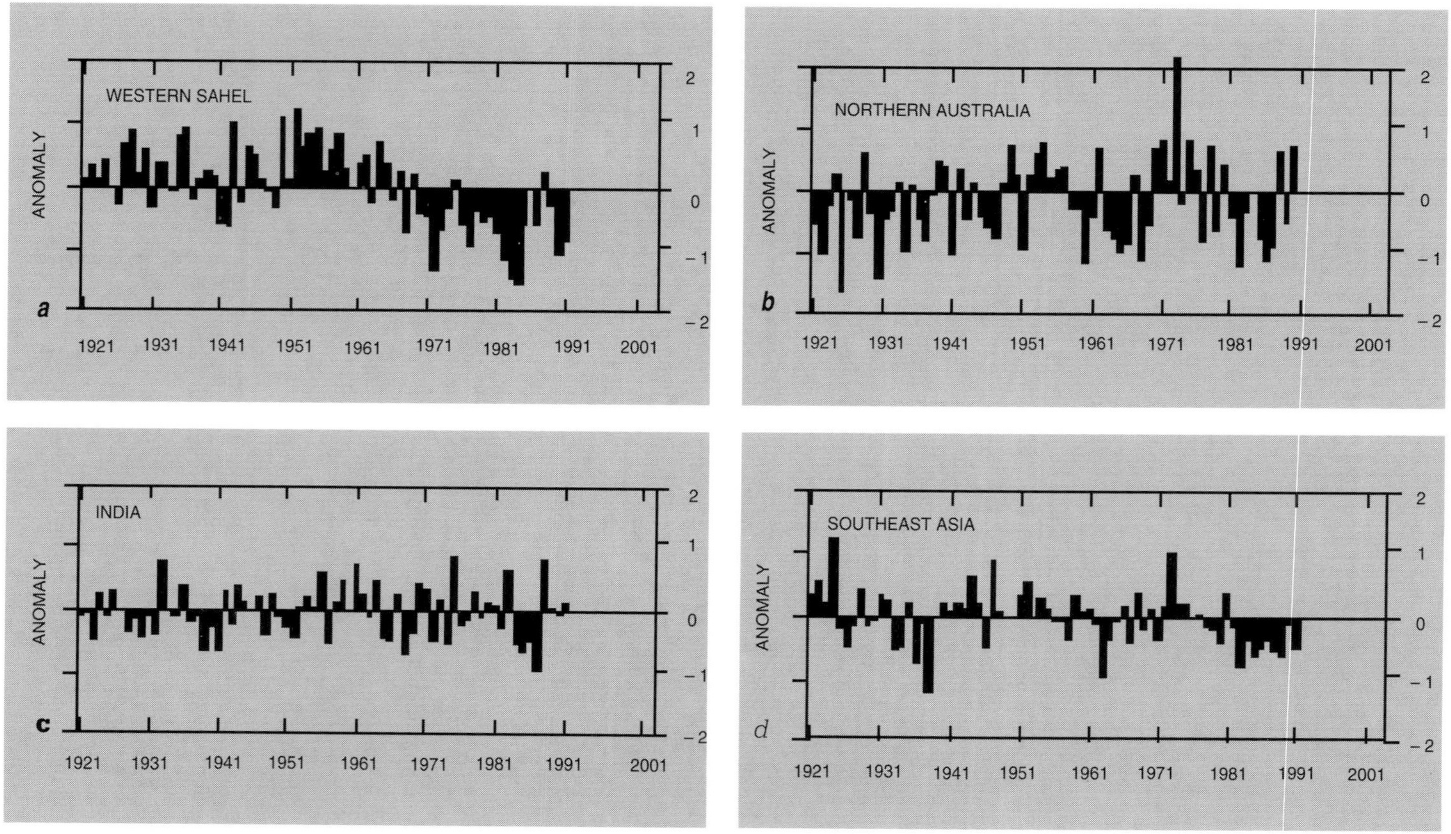

Figure 2 — Precipitation index (average gamma percentiles of station precipitation within the regions) time series for (*a*) western Sahel, June–Sept.; (*b*) northern Australia, Nov.–April; (*c*) India, June–Sept.; and (*d*) south-eastern Asia, April–Sept. (*After Climate Analysis Center, USA*)

Other parts of Africa have also seen severe drought during this period, though few have seen such prolonged desiccation as the Sahel. In Sudan and Ethiopia the drought has in recent years affected quite densely settled agricultural areas. Famine has ensued, leading to massive hunger-relief operations. Many parts of eastern and southern Africa have also suffered severely, though intermittently, especially in the early 1980s and 1990s. The area surrounding the Kalahari and the savannahs of Angola and Botswana have also been affected during part of the period, as has the pastoral land of South Africa and its enclaves.

In South America the major stresses have been in north-eastern Brazil. This unique region, with rainfall in the low-sun season, has a large population of cultivators and pastoralists with a high fertility rate. Drought has repeatedly stressed the region's food system, precisely at a time when pressures were mounting due to the increase in population. Vigorous action by the Brazilian Government, including resettlement programmes, has not avoided great hardships.

Similar stresses have recently afflicted western China and north-west India, regions of monsoon rainfall with high population densities. The Chinese Government has become deeply concerned for the conservation of its extensive lössic soils in the west, the product of millennia of airborne transport of dust and soil particles from arid inner-Asian plateaux and basins. In Pakistan the salinization of much irrigated land has been a major consideration, because most agriculture depends on such land in the desert climate of the Indus-Sutlej basin.

Australia has also suffered a series of drought episodes. The great drought of 1972–1973 was the culmination of prolonged desiccation in many areas; at Alice Springs, for example, the general trend of rainfall was downward from 1947 until 1965, with renewed severe drought occurring in 1972. Heavy rains in 1973–1974 in the east and the interior ended the episode. Severe droughts have also been recorded in 1982–1983 and 1991–1992.

Thus many parts of the world have experienced recurrent droughts. It is also evident from the previous section and from maps A and F of the centerfold that there have been some common features in the years of climate anomalies throughout the world. Many of the recent large climate anomalies have been linked to positive (warm) and negative (cold) phases of the *El Niño*/Southern Oscillation (ENSO*) event. It has been suggested that the severe drought in southern Africa is related to the 1991–1992 ENSO episode.

The droughts which, many generations ago, could have had minimum impacts on society thanks to the use of traditional survival methods are now making havoc of the ecosystems because of the heavy pressure on the

* *El Niño* refers to periodic anomalous warming of the central and eastern Pacific Ocean. The Southern Oscillation is a zonal atmospheric circulation. The two phenomena are closely interlinked, hence the usual reference as ENSO.

environment to satisfy the ever-increasing demand for food, wood, raw materials, energy and water. Nowhere is this more true than in Africa.

A final point concerning recent climatic fluctuations is that the sufferer from drought sees the local landscape shrivelling under a burning sun. He is not to be blamed if he sees the process as wholly local and capable of local solution. He cannot see the gigantic, world-scale mechanisms at work in the atmosphere. Yet much of this problem may spring from these larger influences.

The lesson dictated by this question of scale is that world-scale action is required to solve the problem at the local level. Efforts can be, and are being, made to alter local conditions in a beneficial way — e.g. by adopting less vulnerable forms of food production or better methods of protecting soils and wood supplies. To this end, governments also can take appropriate measures within their competence, such as ensuring that their national observing and communication systems function efficiently. A huge task remains for the World Meteorological Organization and for bodies within the UN and non-governmental organizations: they have to tackle the global component of the changes described above. A co-ordinated world study of climate is needed by every individual nation. In fact, WMO and UNEP have already initiated a project to study and prepare a comprehensive report on the interactions of climate and desertification.

Unfortunately, our understanding of these processes is still too flimsy to allow us to predict future climate, or even next year's weather. Major objectives of the ongoing World Climate Programme, in particular the research component, are: to extend the range of present forecasting techniques, to predict fluctuations lasting decades, and to determine the possibility of lasting climatic change due to changes in atmospheric chemistry. Other programmes are being undertaken under the joint WMO/ICSU/IOC Global Climate Observing System. None of these challenges are easy; some may be impossible, because of climate's inherent unpredictability. All-out efforts are needed to improve our power of prediction.

CHAPTER 5

Causes of climatic variations

To answer the question "What causes climatic variations such as drought, flood, cold/warm spells, etc.?", it is first necessary to make an arbitrary division between microclimate and macroclimate. No real separation of scales exists in nature. The distinction is nevertheless necessary if observers are to understand how climate works.

Microclimate is the word applied to the smallest scale on which we can observe climatic balances, exchanges and processes. Each pasture, each irrigation ditch, each sloping surface has a special climate determined by its local relation to such things as the daily and annual course of the Sun and the prevailing winds and their fluctuations. Microclimates are measurable by means of instruments installed near the ground, or even within the soil. Every farmer is aware of them, knowing for example which of his soils are warmest, driest or most shaded.

Desertification involves a drastic change in such microclimates. If shrubs and trees are felled, for example, the noonday sun will fall directly on hitherto shaded soil; the latter will become warmer and drier, organisms living on or in the soil will move away to avoid the new harshness. The organic litter on the surface — dead leaves and branches, for example — will be quickly oxidized, the carbon dioxide being carried away. So also will be the small store of humus in the soil.

Pastoral use of a soil similarly alters the microclimate (Figure 3). Wandering grazers reduce the grass cover and break up any protective crust that may cover bare soil patches. The pounding of their hooves reduces the fine material to dust, making it erodible by the next strong wind. The soil's porosity is reduced, with the result that the next rainstorm will run off over the surface, eroding much material, instead of percolating to groundwater or recharging soil moisture. A few animals will not do this; a large herd will.

All these changes of microclimate are also ecological changes. The ecosystem is being altered, in most cases adversely. Hence desertification is not only a loss of biological productivity; it is also a degradation of surface microclimates. This has been a vital lesson forced on us by the recent episodes of harsh climate.

Macroclimate is the name given to the larger-scale observable aspects of climate. The word comprehends the ordinary notions of average weather, or the expected range of weather types for a given region. The observed sequence of weather at a locality involves the ceaseless progression past it of differing air masses and disturbances. The rainstorms on which biological productivity depends are carried along by winds that are governed by worldwide energy differences. Their effect is locally observable, but they work in response to larger-scale, invisible processes — hence the word macroclimate.

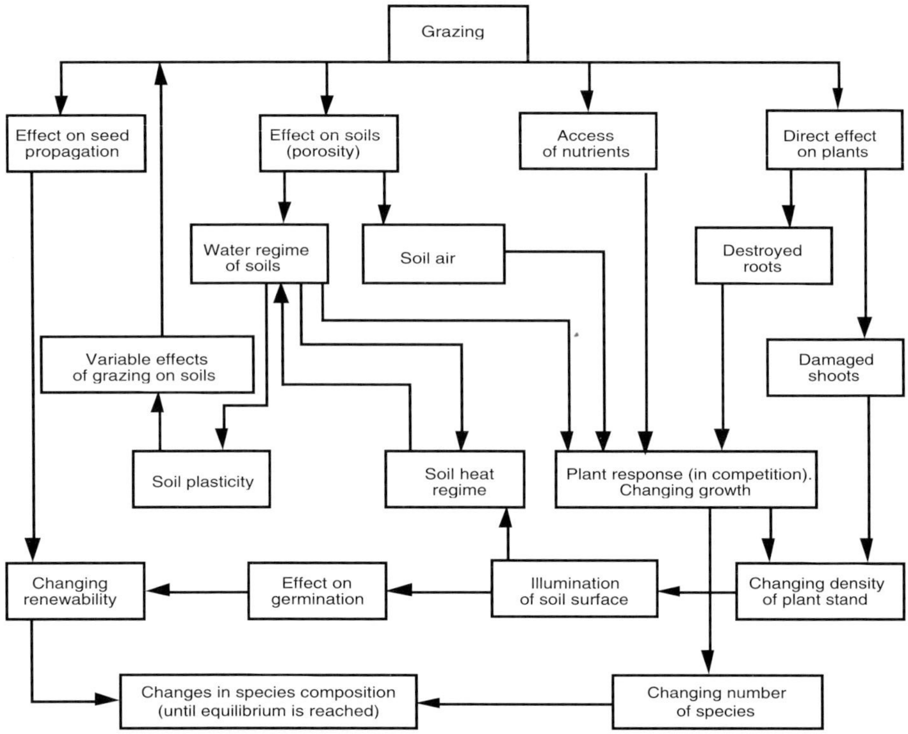

Figure 3 — Schematic description of the alterations in an ecosystem induced by grazing (*After former USSR Commission for UNEP*)

The general circulation of the atmosphere and ocean — the name we give to their large-scale motion — controls the macroclimate. The general circulation in turn is governed by the exchanges of energy between Earth, Sun and outer space. It undergoes certain imperfectly understood oscillations that have profound influences on local climate. For example:

(*a*) The principal rainstorms of sub-Saharan Africa are actually controlled by the Inter-Tropical Convergence Zone (zone of convergence of inter-hemispheric tropical winds) and the westward-moving disturbances in the easterly current that overlies the shallow south-west monsoon — the current that brings the moisture into the continent. We do not know why these disturbances have in recent years been less effective in promoting rainfall. The answer probably lies with a change in the general circulation having little to do with Africa itself.

(*b*) At the same time as the weakening of the above systems, there have been some anomalies in the space-time characteristics of the other rain-bringing systems such as the cyclones, jet streams, easterly/westerly winds, extratropical weather systems, etc.

(*c*) *El Niño* and the Southern Oscillation are two interrelated systems which have been widely teleconnected with many climate anomalies throughout the world. Other climate anomalies have been associated with westerly/easterly phases of the tropical stratospheric winds (quasi-biennial oscillation), and with intraseasonal wave and other climate systems induced by large-scale anomalies in the general circulation.

CHAPTER 6

Climatic feedbacks: Does human interference worsen drought?

Since the impact of drought has been so widespread in the past two to three decades, it is natural to ask the questions: Does human interference worsen drought? Does desertification create feedbacks that exacerbate the climatic stresses? Can altered ways of using land and water put matters right?

Of the feedbacks that have been defined, the most thoroughly investigated is that of surface albedo — the reflectivity of soils, vegetation and structures. Desert surfaces are visibly reflective: a day-time journey across them is hazardous unless one's eyes are protected against the glare. Figure 4 illustrates the point: even from an orbiting satellite the Sahara and the desert surfaces of the Near East appear surprisingly bright. The brightest are the sandy deserts, which reflect back more than 30 per cent of the sunlight and perhaps more than 50 per cent in some areas.

It follows that desertification should tend to increase surface albedo, since a plant cover is usually darker-coloured than bare soil, sand or rock, and hence absorbs more solar radiation. Areas of desertification can actually be detected in satellite images by their lighter colour. Patches of this kind can be seen in satellite and aerial pictures of the Sahel, Madagascar, Lesotho and the North American Great Plains. Qualitatively, desertification can make the Earth a slightly brighter planet when viewed from space.

In the mid-'seventies it was discovered, moreover, that some subtropical areas, chiefly the Sahara-Arabian Peninsula belt, actually lost more energy to space than they received from the Sun. These areas actually appeared, even in summer, to be zones of energy loss to space. The reason was twofold: the large reflection of solar radiation from the bright desert surfaces, and the ease with which longer-wave terrestrial radiation could escape upwards. The subtropical desert areas were revealed as energy sinks for the Earth atmosphere system.

In spite of this unexpected observation, it is also a fact that the subtropical zone is very warm at the surface and up through the troposphere (the bottom 15 km). How is this warmth maintained, if the larger desert areas lose heat to space? The answer is: general subsidence of the air above the deserts. There is sinking motion in the middle and upper troposphere (i.e. from about 4 to 12 km). The descending air is warmed and rendered very dry by compression. This warmth and dryness, together with the overall sinking motion of the atmosphere, suppress clouds and rainfall, and make the Sahara and parts of south-west Asia nearly rainless.

Figure 4 — METEOSAT image of Africa, showing the highly reflective surface of the Sahara and other desert areas (*Photo by courtesy of the Meteorological Service of France*)

The idea that raised albedo should increase the heat loss, and hence the rate of subsidence, was put forward in 1975 by J. Charney, who in turn owed the idea to a suggestion made earlier on a more local scale by J. Otterman. Charney's idea has subsequently been tested by several large-scale experiments, in which numerical models of the general circulation have been run with artificially high albedos over the desert areas. These experiments have confirmed that large increases in the albedo of subtropical areas should reduce rainfall, but the quantitative effects in actual situations have not yet been firmly established. Some of the models also suggest changes of rainfall in other areas, i.e. teleconnections. It may be the case, therefore, that the droughts afflicting the arid zone are part of a much larger pattern of anomalous climate that we have not connected up.

Map A

1972 DROUGHTS
March - December

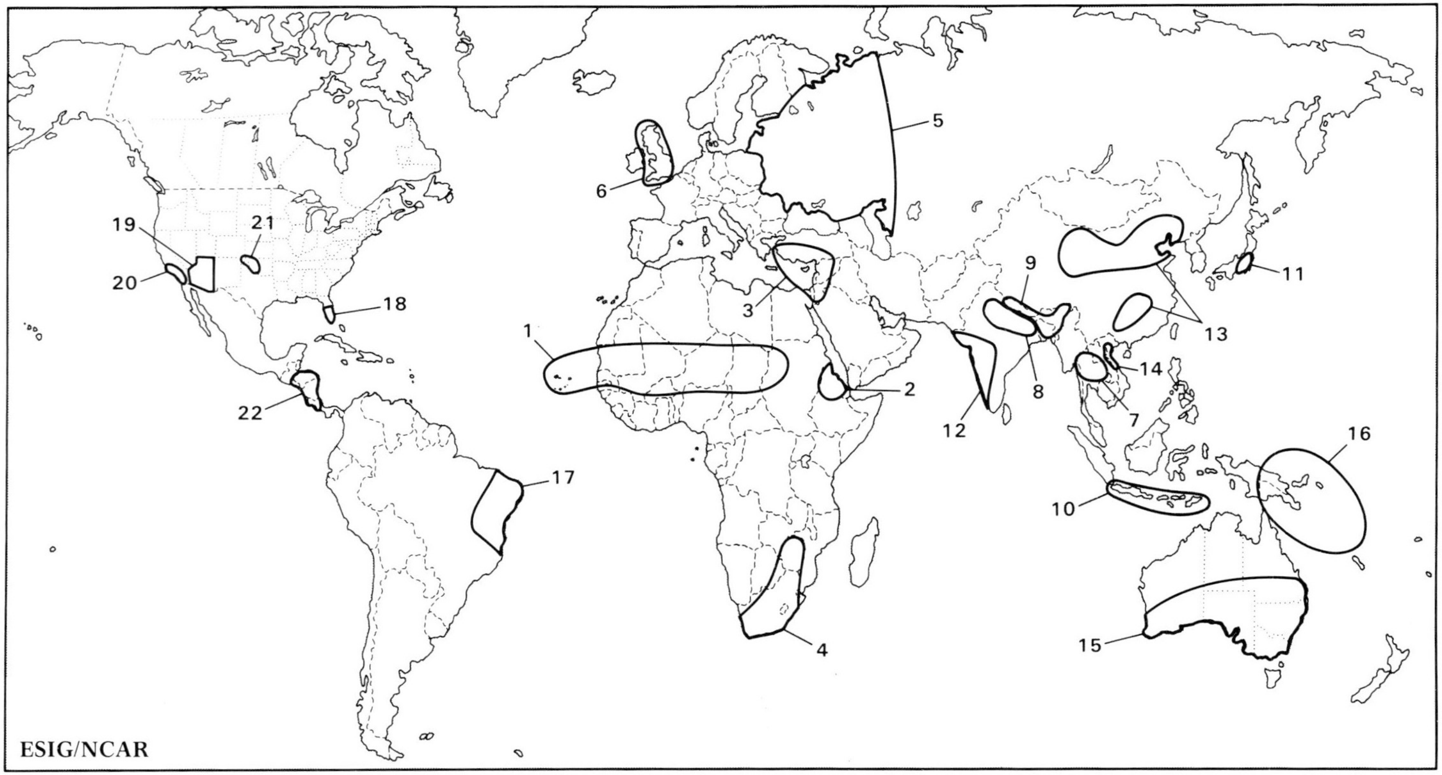

Legend to Maps A, B, E and F: World patterns of rainfall anomalies during the ENSO events of 1972–1973 and 1982–1983

Africa and Middle East

1. African Sahel: Mali, Mauritania, Senegal, Cape Verde Islands, Niger, Upper Volta, Chad, W Sudan - Mar. thru Dec.
   a. Drought in Sahelian region that began in 1968 intensified in 1972, crop failures, livestock deaths, water shortages, famine
   b. Chad: 1972 annual rainfall lowest since 1943
   c. Agades, Niger: 1972 rainfall total 30 mm vs 164 mm average
2. N and C Ethiopia: Tigre, Wollo, Eritrea Showa provinces - Mar. thru Dec.
   a. Drought began in 1971 extensive crop losses, 80% of cattle lost, famine
3. Mediterranean Middle East: Jordan Valley, Syria, Turkey, Israel, Lebanon, Cyprus - Sept. thru Dec.
   a. Drought said to be worst of century, water shortages in many urban areas, livestock losses, crop damage, no rainfall in some areas during all of 1972
4. S Africa: Zimbabwe (Rhodesia), Republic of S Africa - Nov. thru Dec.
   a. Failure of summer rains, crop and livestock losses

Europe and USSR

5. USSR: European areas - May thru Sept.
   a. Worst drought in century: less than 25% of normal rainfall in July and Aug. over large area
   b. Mid-July thru Aug. heat wave in NE USSR caused fatalities, crop damage, forest and peat fires (over large areas of E Russia)
      (i) Hottest summer on record in Moscow, heat wave extended into N Finland
   c. Unusually cold and dry winter also hurt wheat crop, spring also dry
6. United Kingdom: Britain - July thru Dec.
   a. Many locations reported 1972 annual precip. as lowest since 1921
   b. Water shortages in many areas, esp. Scotland and E areas, crop damage Asia
7. Thailand - Apr. thru June
   a. Drought during beginning and middle of rainy season, crop and livestock losses
8. N India: W Bengal, Uttar Pradesh, Bihar - May thru early June
   a. Heat wave and drought, temps. consistently over 100°F, death toll over 600, New Delhi severely affected, damage to crops, livestock
9. NE India, Nepal, Bangladesh - June thru Aug.
   a. Failure of summer monsoon
   b. Assam province, India received 855 mm rain in July compared with 2855 mm average
10. Indonesia: Java, Madura, Bali and SE areas - June thru Oct.
11. Japan: Tokyo area - June thru early July
   a. Drought, worst water shortage in area since 1964
12. W coastal India - July thru Dec.
   a. Severe monsoon drought in 4 western states along coast, esp. Maharashtra, crop damage
13. China - July thru Dec.
   a. Widespread drought, crop damage, livestock losses, water shortages
14. N Vietnam: 5 provinces s of Hanoi - Nov. thru Dec.

Australia and Oceania

15. S Australia - Mar thru Jan 1973
   a. Severe drought; SW W Australia in 3rd yr of dry weather, winter 1972 rains below average for most areas in S Australia, conditions worsened as drought extended into summer rainfall areas of SE, drought ended late 1973 and into Feb. with heavy rainfall
   b. Affected areas included: SW W Australia, S Australia, New South Wales, S Queensland, S Northern Territories, Victoria (esp. Victoria)
   c. Heat waves
      (i) May 1972: Sydney and neighbouring areas, bushfires in New South Wales and Victoria
      (ii) Dec. 1972: SE areas, esp. Victoria, persistent heat heightened effects of drought
      (iii) Jan. 20-23, 1973: Melbourne areas, fatalities
16. Papua New Guinea and Melanesian Islands - Apr. thru Aug.

USA, Canada, Latin America and the Caribbean

17. NE Brazil: esp. Ceara, Bahia, Piaui - Mar. thru Dec.
   a. Effects of drought began to be felt in Apr., severe in some locations by Aug.
   b. Area experienced very severe drought in 1970-71
18. SE USA: S Florida - Mar. thru Nov.
   a. Reported in Nov. that previous 29 months had below-normal precip., crop losses, Everglades "threatened"
19. SW USA: Arizona - Mar. thru May
   a. Jan. thru Apr. in Phoenix and Tucson completely rainless, driest 4 months on record
   b. Extended drought on Navajo reservation, esp. severe by end of May
20. W. USA: S California - Mar. thru Dec.
   a. Los Angeles area rainless from Mar. 28 thru end of year
   b. San Diego, 1st 4 months of 1972, Apr. in particular, also very dry in other parts of S and W USA: El Paso, 90 consecutive days of no rain reported at end of Apr.; Grand Junction, Colo., driest Jan. thru Apr. on record; Cheyenne, Wyo., Apr. was 6th consecutive very dry month; Charleston, S. Ca., driest Apr. on record; Macon, Ga., 2nd driest Apr. on record
21. USA: W Oklahoma - Mar. thru May
   a. Worst drought in W Oklahoma since 1930s, loss of 25% of wheat crop, reported as 2nd yr of drought in area
22. Central America: Honduras, Nicaragua, Costa Rica - June thru Dec.
   a. Abnormally light rains during summer rainy season, crop damage
   b. Drought also reported in El Salvador during June and July
   c. Nicaragua reported severe 1972 crop losses: 80% corn, 35% beans and rice, 20% wheat (reported in Jan. 1973)

1972 FLOODS/HEAVY RAINS/SEVERE SUMMER STORMS

March - December

Map B

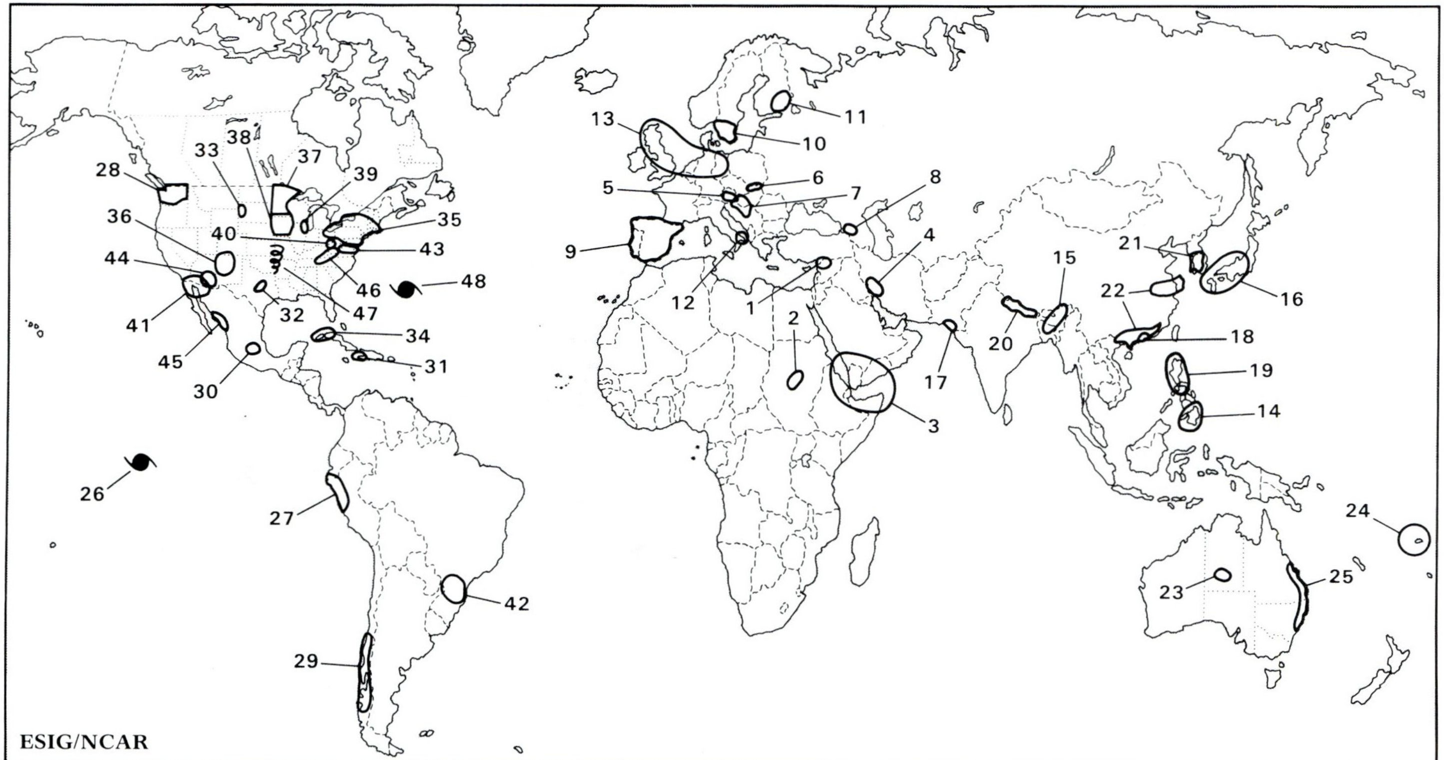

Asia

17. NE Pakistan, N and NW India: Punjab, W and E Uttar Pradesh - Feb. thru June
18. NW Pakistan - Feb. 7-20
19. S China: Guangdong, Guangxi - mid Feb. thru mid-Aug.
    - *a.* Periods of torrential rains, high winds, extensive flooding, property and crop damage, fatalities
    - *b.* Esp. May 2-15, June 27 thru July 10, Tropical Storm Dot Aug. 15-18
    - *c.* May totals for S China 200-600% normal; worst flooding in 30 yrs in SE Guangdong
20. S and C Philippines - Mar. 20-29
    - *a.* Typhoons, flooding, high winds, property and crop damage, fatalities
    - *b.* Typhoon Mamie: Mindanao - Mar. 20
    - *c.* Typhoon Nelson: Visayas - Mar. 23-29
21. Sri Lanka - May; late Nov. thru late Dec.
    - *a.* Esp. Nov. 21-27 and Dec. 19
22. SW coastal Burma - May 4
    - *a.* Cyclone, crop and property damage, fatalities
23. Hong Kong - late May
    - *a.* Torrential rains, landslides, fatalities
    - *b.* Esp. May 28-29
24. C and E China: Hunan, Jiangxi, Zhejiang, Jujian, Sichuan - June
    - *a.* Extensive flooding, property and crop damage, fatalities
    - *b.* Esp. S Fujian, worst flooding in 30 yrs
25. Coastal W India - June
    - *a.* Coastal flooding from heavy monsoonal rains
26. N and C Philippines: Luzon and central islands - June thru Oct.
    - *a.* Severe typhoon season, periods of severe flooding, crop and property damage, fatalities
    - *b.* Esp. Typhoon Nancy - Oct. 12-16
27. Indonesia: S Sumatra - June 3-5
    - *a.* Heaviest monsoons in 75 yrs, severe flooding, property damage and fatalities

43. E Australia: S coastal Queensland - Sept. 20
44. E Australia: coastal New South Wales - Oct. 9-11
45. Pacific hurricanes and tropical storms - Jan. thru Dec. 1982
    - *a.* 19 named storms (14.4 is 16-yr average)
    - *b.* 11 were hurricanes, 8 tropical storms (plus 7 tropical depressions)
    - *c.* One of the most active seasons on record

USA, Canada, Latin America and the Caribbean

46. Coastal Brazil: Rio de Janeiro - Jan. 5
    - *a.* Heavy rains, floods, landslides, fatalities
47. SW Colombia - early Jan.
    - *a.* Torrential rains, flooding, landslides, esp. Jan. 4 and 8
    - *b.* Esp. Nanino province and Manizales area
48. W USA: California - Jan. thru Apr.; Sept.; Dec.
    - *a.* Periods of torrential rains and severe storms, high winds, flooding, mud and landslides, property and crop damage, fatalities
    - *b.* San Francisco region - Jan. thru Apr. (esp. Jan. 1-6)
    - *c.* Entire San Joaquin Valley - Sept. (esp. Sept. 23-25)
    - *d.* San Francisco region into N and C areas - Dec.
49. s Ecuador, N Peru - mid Jan. thru July
    - *a.* Heavy rains, flooding, river flooding, extensive property damage, fatalities
    - *b.* Esp. Jan. 23-25, Feb. 20
50. NW US: NW Washington, esp. Bellingham area - Feb. 13-14 and 23-24
    - *a.* Severe storms, heavy rains, snowmelt
51. N, E and C Bolivia - Mar.
    - *a.* Heavy rains, flooding, landslides, crop damage, fatalities
    - *b.* Esp. Mar. 2-5 in Beni and Mamore River valleys
    - *c.* Esp. mid-Mar. in E Bolivia Santa Cruz area
52. N USA: N Indiana, S Michigan, NW Ohio - Mar. 14-22
    - *a.* Rain, snowmelt, flooding
53. SE Peru: esp. Cuzco - Mar. 23-26
    - *a.* Heavy rains, landslides, floods, fatalities

67. C USA: Missouri, Kansas City area - Aug.
    - *a.* Esp. Aug. 12-13
68. C Canada: C Saskatchewan - Aug.
    - *a.* Hail, heavy rain, crop damage
    - *b.* Near Prince albert
69. SE USA: C and E Tennessee - late Aug.
70. W US: Utah, Salt Lake City area - Sept.
    - *a.* Wettest in 76 yrs
71. Coastal Guatemala, El Salvador - Sept. 17-22
    - *a.* Hurricane Paul, heavy rains, floods, large landslides, property and crop damage, esp. in El Salvador
72. N Mexico: Sinaloa - Sept. 29-30
    - *a.* Hurricane Paul, heavy rains, high winds, flooding, property and crop damage
73. C Chile: Temuco region - Sept. 30
74. Coastal Ecuador, N Peru - Nov. thru Dec. 1975. USA: W Hawaiian Islands - Nov. 24
    - *a.* Hurricane Iwa
    - *b.* Kauai, Oahu Islands
76. W Mexico: Manzanillo area - Nov. 25-26
77. N Bolivia: Baures River basin - Dec.
78. USA: Mississippi Valley - Dec.
    - *a.* Illinois, Arkansas, Missouri, early Dec.
    - *b.* Louisiana, Mississippi - late Dec.
    - c. Illinois, Iowa, Missouri, Louisiana, Mississippi, Arkansas had wettest Dec. on record
79. Tornados: USA - Jan. thru Dec. 1982
    - *a.* Total of 1047 (30 yr mean is 739)
    - *b.* 23.6% strong (average is 33%), .5% were violent (average is 2%)
    - *c.* Mar. thru May 1982 had highest frequency of tornados on record, 365 (previous record 275)
      (i) Damage concentrated in S and C states in Mar.; S Mississippi Valley in Apr.; S Illinois in May
80. Atlantic hurricanes - Jan. thru Dec. 1982
    - *a.* 2 named hurricanes
    - *b.* Least active season in more than 50 years (usual is 10 named storms with 6 becoming hurricanes)

Map F

1982 DROUGHTS
January -December

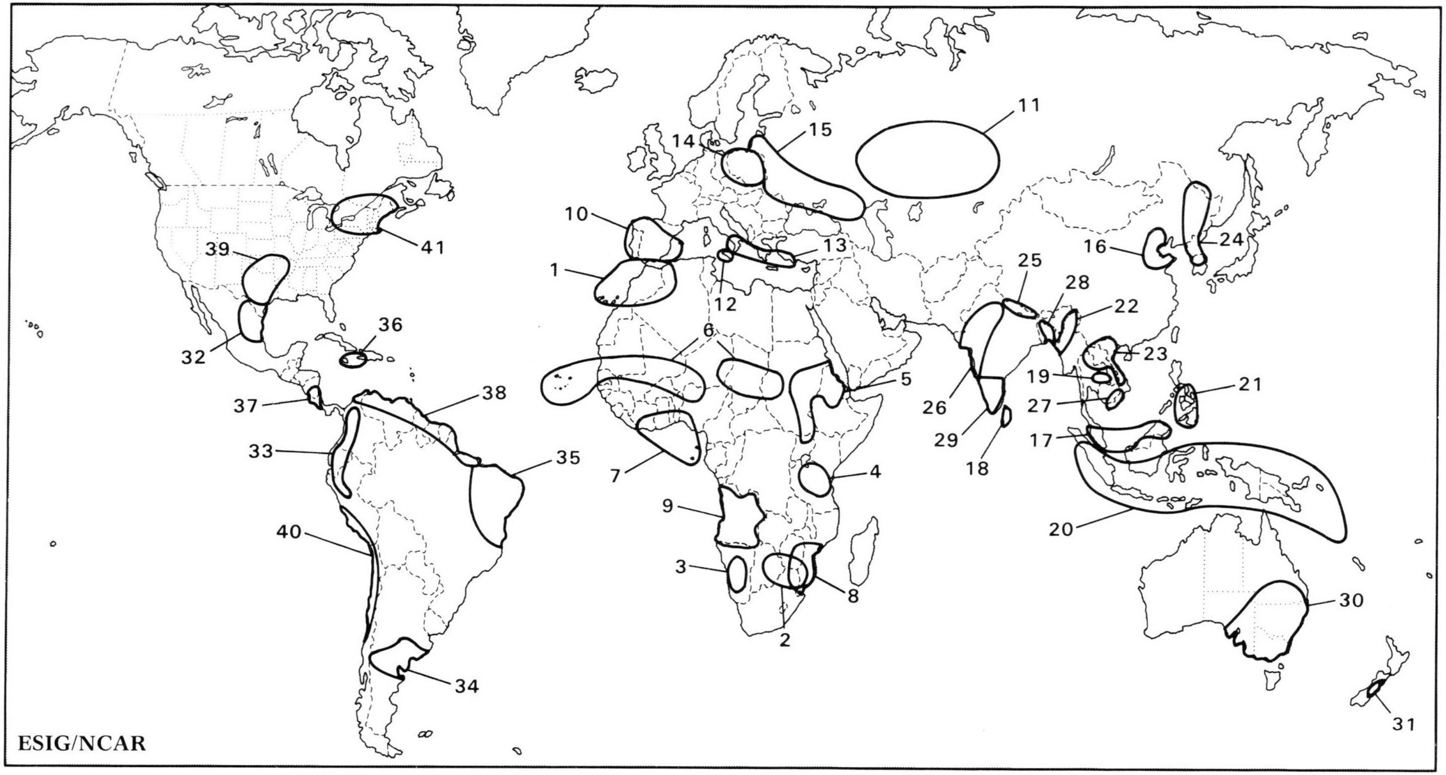

Africa and Middle East

1. NW Africa: W Algeria, Morocco, Canary Islands - Jan. thru Mar. or Apr.; Morocco - Nov. thru Dec.
2. SE Africa: E Botswana, S Zimbabwe, S Mozambique, N Republic of S. Africa - Jan. thru Apr. or May
3. C Namibia - Jan. thru Apr.
4. N Tanzania, E Rwanda - Mar. thru June
5. NE Africa: E Sudan, NE Uganda, N Ethiopia, Gondor, Wollo, Eritrea, Tigrai - June thru Aug.
6. African Sahel: Cape Verde Islands, coastal and N Senegal, N Gambia, S Mauritania, C Mali, N Upper Volta, W Niger, N Nigeria, C Chad, C Sudan - June thru Oct. (considerable variability throughout period)
7. W coast Africa: Ivory Coast, Ghana, Togo, Benin, Nigeria, Sao Tome, Principe Island - Aug. thru Dec.
8. SE Africa: E Zimbabwe, S Mozambique, NE Republic of S. Africa - Sept. thru Dec.
9. Angola - no dates or exact locations available

Europe and USSR

10. Spain, Portugal - Mar. thru Apr.; July thru Aug.; Dec.
    a. Widespread and scattered drought continued from 1981, heat waves, forest fires, crop damage, water rationing
    b. Dec only 10% of normal precip.
11. C USSR: W Kazakhstan, S Urals, W Siberia - Apr. thru Aug.
12. S Italy: Sicily - May thru July
13. Mediterranean states: S Italy, Greece, W Turkey - late June
    a. Heat-wave-related deaths, forest fires, crop losses
14. Poland - July thru Nov.
    a. Also parts of Yugoslavia and E Germany
15. SW USSR: N Caucasus, Ukraine, Baltic states - Sept. thru Nov.
    a. i.e., Krasnodar only 12-40% normal precip. for period

Asia

16. N China plain: Shandong and Hebei - Jan. thru May
17. Malaysia - Jan. thru Apr.; E Sabah - thru May
18. Sri Lanka - Jan. thru Apr.; mid-June thru Oct.
19. SE Thailand - Jan. thru Mar.
20. Indonesia, Papua New Guinea, Melanesian Islands - Mar. thru Dec.
    a. Beginning in Indonesian archipelago and N Sulawesi, spreading to entire area by July; worst drought in 10 yrs
21. S and C Philippines - Mar. thru May; mid Sept. thru Dec.
22. S Burma - May thru Oct.
23. Laos, N Thailand N and C Vietnam parts of Kampuchea and S Vietnam - May thru Aug., C Vietnam - thru Dec.
24. N and S Korea, NE China - June thru July
25. Nepal - June thru Dec.
    a. Failure of summer monsoon, severe drought
26. W to N India: Uttar Pradesh, Rajasthan, Gujarat, Maharastra - July thru Sept.
27. S Vietnam, E Kampuchea - July thru Aug.
28. NE and coastal Bangladesh - Aug. thru Nov.
29. S India: Kerala, Tamil Nadu - Aug. thru Dec.

Australia and Oceania

30. SE Australia - Apr. thru Dec.
31. New Zealand: E coastal South Island, Canterbury Plains - June thru Aug.
    a. Dry in New Zealand entire year except S and W South Islands

USA, Canada, Latin America and the Caribbean

32. C and NE Mexico, SE Texas - Feb. thru Mar.; June thru Nov.
    a. i.e., NE Mexico 20-40% of normal in Feb. and Mar.
33. N Peru, Ecuador, Colombia: Cauca and Magdalena River valleys - Feb. thru Dec.
34. Argentina: Buenos Aires province - Mar. thru May
35. NE Brazil - Mar. thru July; Sept. thru Dec.
36. S Haiti and Jamaica - Mar. thru June; Sept. thru Dec.
37. Costa Rica, S Nicaragua - July thru Aug.
38. NW Colombia, N Venezuela, Guianas - Sept. thru Dec.
39. S USA: Texas, parts of Oklahoma, Arkansas - Sept. thru Oct.
40. W coast S America: S Peru, Chile - Nov. thru Dec.
    a. Heat wave, dry
    b. Late Dec., many heat-related deaths in Lima
41. NE USA and SE Canada: S Ontario and Quebec - Nov. thru Dec.
    a. Deficient snows, ski areas hurt
    b. Mt. Washington, New Hampshire – no snow cover on Dec. 25, 1st time in 50 yrs
    c. N Maine, lowest snow in Dec. on record
    d. At least a dozen major cities in east had one of the warmest Dec. on record, i.e., Dec. 25 warmest in 142 yrs in Toronto

A second possible feedback whereby aridity tends to reinforce itself, and hence drought to beget drought, is diminished water storage within continental areas, especially Africa. At points far from an ocean, e.g. the centre of the Amazon basin, it is believed that much rainfall actually derives from locally evaporated water, and not from moist air from the ocean. The same situation is true for the Zaire/Congo basin and parts of inland Africa.

In the region worst affected by drought, the sub-Saharan belt, the initial source of rainfall is the south-west monsoon blowing in across the Guinea coast. Far inland, however, the rain that falls will very probably already have been precipitated once or more, and re-evaporated.

If the water-holding capacity of the soil is reduced (because of the loss of fine mineral and organic materials), there will be increased resistance to such re-evaporation, and decreased reservoir from which to draw it. Removal of deep-rooted shrubs and trees also decreases the opportunity for transpiration. Deforestation and desertification should accordingly reduce the opportunity for rainfall of this recycled variety. Numerical model tests have indicated that the rain falling in a specific rainy season depends greatly on the moisture content of the soil at the outset.

A third possible feedback mechanism which has also been investigated concerns surface roughness. Most continental surfaces are anything but smooth and exert a drag on winds moving across them. The drag is actually exerted by the roughness elements projecting above the surface, such as grass spikes, trees or buildings. It dissipates some of the wind's energy, and also leads to an exchange between wind and ground.

A desert surface is much smoother than, for example, a forest. It has been suggested (by Y. C. Sud and others in the United States) that desertification reduces the roughness of the surface, by eliminating tall trees and shrubs. Numerical modelling indicates that such a change towards a smoother surface should diminish rainfall in some areas, and increase it elsewhere. The effect is not simple, but does appear to be large.

All three of these feedback mechanisms are at work along the margins of the world's deserts. It is not certain, however, that they have been effective in prolonging or intensifying drought. All the modelling exercises carried out to date have had to assume much-exaggerated changes, because the available models are too crude to respond to the kinds of small, gradual changes that are actually taking place.

The above work has demonstrated qualitatively that processes that may amplify the effects of desertification do exist. The changes being wrought by human land use may actually contribute to the worsening of climate. The tests so far carried out support this view. But most climatologists believe that the fluctuation in the macroclimate discussed above is part of the internal variability of the climatic system. They consider that normal or even humid conditions will return in due course to the drought-stricken areas, as the atmospheric circulation rights itself. There is at present no way of deciding whether the climatologists are right or wrong, or whether the changes in Africa are due to a permanent deterioration, caused mainly by human interference.

What is unmistakable, however, is that there have already been long-lasting changes in surface microclimates. Woody vegetation, stripped off so much of the arid lands to provide fuel or fodder, will not regrow quickly. Reserves of shallow groundwater must have been heavily depleted in many areas. Damage to soils has been severe. These remarks apply not only to Africa, admittedly the worst case, but to all those parts of the arid zone where drought has occurred or been persistent in areas of rapid population increase.

CHAPTER 7

Climate and the control of desertification

In what ways can a knowledge of climatic processes help combat desertification? We have already seen that climatic variation cannot yet be predicted with any certainty. Are there other roles that the climatologist can play?

WMO gave general answers to these questions in its action plan to combat desertification, adopted in 1978 at the thirtieth session of the Executive Council, in support of the UN's Plan of Action to Combat Desertification. Since that time WMO and its specialists have done what they can to help in the design and execution of countermeasures. The plan was reviewed in 1990 by the Council at its forty-second session, which drew up a new strategy for WMO activities in support of the combat against drought and desertification. The strategy was recently updated by an informal expert group meeting held in WMO during November 1992. An abridged version is attached as Annex I.

Much of this effort is directed at training nationals in the application of meteorological and hydrological data to avoid or assess land degradation and desertification through soil loss, dust and sand transport, in management of natural pastures and in irrigation scheduling in order to ensure sustainable use of land and water resources.

What general principles underlie the advice that WMO has given?

One cardinal principle on which the Organization insists is that proper monitoring of climate itself, and of associated hydrological and ecological processes, is imperative. Governments, private organizations and individuals must recognize that patient, long-term, consistent observations are absolutely central to the battle to restrict desertification. One cannot fight an enemy effectively unless one has taken its measure.

Many parts of the arid zone have low incomes, and the governments involved have difficulty maintaining adequate services. There have been many closures of weather stations in the past two decades as finances have become tighter. To discontinue the record of a station opened decades ago is to rob the climatologist of proof that climate is changing and of the details of that change. Far from losing such records, the climatologist needs more stations, with more sophisticated observing capacity. This point needs to be hammered home. *More* records are needed, not less.

The methods of such monitoring are improving by leaps and bounds, thanks to electronic technology. Satellite observation allows continuous monitoring of various climatic elements that have hitherto been hard to observe. Among these, for example, is cloudiness. The cloud systems of the rainy season can now be tracked across the arid zone in a graphic fashion. Many properties of the troposphere and stratosphere can also be monitored by satellite techniques. All countries in the arid zone need access to this vital information.

Of special interest is the recent development of techniques to measure vegetational change, essentially by means of the varying amounts of green pigments in the leaf canopy, using routine satellite observations. This has made it possible, for example, to observe directly the northward and southward migration of the rainy season in Africa, which affects the greenness of the surface. The same techniques can be used to detect the influences of drought in retarding the advance of the growing season and to demonstrate the impact of desertification and deforestation. The degree of resolution is not, of course, adequate for detailed observation, but much valuable information is now available on the continental and regional scales.

It would be a mistake, however, to infer that the availability of satellite information obviates the need for "ground truth", i.e for the continued observation of processes and events of ground level. The satellite record enriches and diversifies the resources with which desertification can be assessed and fought; it does not reduce the value of ground observation, and certainly does not replace ground truth data.

A second fundamental principle is that land-use control is the key to improving climates and to minimizing the effects of drought. It has been shown above that the surface microclimates are adversely affected by desertification. Measures to control the latter do not merely protect soil and vegetation, but also improve the microclimate from the point of view of biological productivity. To quote from a revised edition of the report prepared under WMO auspices for the UN Conference on Desertification:

> . . . land-use control is also microclimate control. Good land-use practice aims to enhance the biological productivity of the land surfaces, i.e to restore productive ecosystems. Amelioration of surface microclimate is best achieved in this fashion. This does not mean that climatologists have nothing to offer in a direct form to the combat against desertification. It does mean, however, that their input is most effective if it is made in the light of a full understanding of ecosystem processes, under both managed and natural conditions.

The political and managerial difficulty of achieving such control is formidable. It has been the main object of study of many workshops in many countries. It is currently on the agenda of the UN bodies, the World Bank and many other regional and international organizations.

A third principle underlying WMO's advice and action is that all countries in the arid zone should provide advisory and technical services that make use of historical and current information about weather and climate. A main component of the World Climate Programme deals with meteorological applications. In addition, the WMO Agricultural Meteorology Programme aims at promoting the application of meteorological and hydrological data to agriculture, land use, and crop production and protection in a sustainable manner, without causing land degradation. Much current and historical information is available in official records. In times of climatic stress (as well as in all long-range planning exercises) it is urgent that this information be used and made widely available. Moreover, research institutions should be aware of the availability of such material and should make good use of it.

The special problems of the Sahel have been addressed in this fashion by the co-operative AGRHYMET programme, which involves many of the drought-afflicted countries. The programme, which has its regional centre at Niamey, in Niger, is engaged in extensive training, information dissemination, crop assessment and forecasting, and various kinds of agronomic development, all in respect of drought-stressed agriculture. Its major objective is to render as useful as possible the resources of agrometeorology and hydrology, especially in relation to food production. Both FAO and US National Oceanographic and Atmospheric Administration crop-assessment models are used, and early warning systems of potential shortfalls of food production have been devised for Member States. Pilot projects in Mali and elsewhere in the sub-region are being monitored. AGRHYMET (Figure 5) is a model of its kind, but still has a long way to go in coping with an extremely difficult situation — and especially in reaching the individual farmer. Other recent efforts by WMO include the establishment of the African Centre of Meteorological Applications for Development (ACMAD) in Niamey jointly with the Economic Commission for Africa (ECA), and the joint UNDP/WMO Drought Monitoring Centres for Eastern and Southern Africa in Nairobi (Kenya) and Harare (Zimbabwe). These three centres must be supported if the adverse effects of drought and desertification are to be avoided.

A fourth position maintained by WMO has been its insistence on the continuing need for research. In times of crisis people often say: "The time for research is past. We have plenty of knowledge; the job now is to apply it." Nothing could be farther from the truth, and nothing more damaging to the fight against desertification.

A major need, for example, is for more detailed study of the albedo, water storage and roughness feedbacks, as demonstrated by modelling exercises. More generally, we need to know how the tropical and subtropical climate system, especially in the monsoon areas, actually works. What is the nature of the huge planetary perturbations now being revealed, such as *El Niño* and the Southern Oscillation? How do such perturbations, and other dynamical phenomena, relate to the prolonged drought episodes afflicting so much of the arid zone? The work of explaining the mechanics of tropical and subtropical climates has hardly begun as yet and a huge job awaits the attention of the dynamical modellers. One of the objectives of ACMAD, when it becomes fully operational, will be to provide answers to the above questions as far as Africa is concerned.

Figure 5 — AGRHYMET Centre (Niamey) (*WMO photo*)

At the other end of the scale, extensive research is needed into the microclimatological relationships of vegetational changes in the arid zone. What effects do such changes have on albedo, water-exchange and storage capacity, and on the carbon cycle? WMO collaborates with many national and regional agricultural research institutes in promoting agrometeorological research in the arid zone.

Finally, the Organization has considered it necessary to warn the world community against certain suggested remedies. Flooding of inland desert basins, for example, is not likely to increase regional rainfall. Tree-planting, although often beneficial to the soil and nearby cropped areas, will not significantly alter the macroclimate. In few parts of the arid zone does it seem likely that weather modification, chiefly cloud-seeding, will be effective. It is natural that nations and organizations investigate such schemes, but professional opinion is that they are most unlikely to be effective — and may even worsen matters.

CHAPTER 8

Climate change and the future

Do desertification and other human activities play any role?

Climate change and desertification have become some of the most pressing environmental issues in the recent years. The United Nations organizations and many international institutions have created a number of special programmes to deal with these issues. The United Nations efforts have included the enacting of several conventions to protect the natural environment from any climate and ecological changes. One can mention the Framework Convention on Climate Change, which has been signed by over 150 countries, the Biodiversity Convention and the Forestry Principles Statement which emanated from the Rio Conference. Other international efforts have included the launching of the Intergovernmental Panel on Climate Change by WMO and UNEP; the Global Climate Observing System (GCOS) of WMO, the International Council of Scientific Unions (ICSU) and the International Oceanographic Commission (IOC); the Desertification Control Programme Activity Centre (DC/PAC) and Oceans and Coastal Areas Programme Activity Centre (OCA/PAC) of UNEP; the Global Sea Level Monitoring System of IOC; the International Geosphere-Biosphere Programme of ICSU; and many others. WMO is actively participating in most of these programmes.

WMO activities are described in Annex I.

The key questions which many of these efforts have attempted to answer are:

- What is the future of climate at global, regional and local levels?
- Is climate change a reality or a matter of fuss?
- What are the potential causes of any climate change?
- Can desertification and other human activities change climate?
- How do the natural forces and human activities interact during the climate-change processes?
- What are the potential impacts of any climate-change scenarios and the corresponding mitigation strategies?

Although the details of future climate cannot be accurately predicted, certain broad trends are considered probable as a result of altered atmospheric compositions.

The most universal and widely accepted opinions on climate change and future climate have been provided by the three working groups of the IPCC on Scientific Assessment, Impact and Response Strategies. The potential causes of climate changes have been classified as natural or human-induced. The natural causes can result in permanent changes in the natural characteristics of the climate-controlling systems. Some of these systems have been discussed in the text. The human factors which have the potential to change the natural climate include desertification, air and water pollution, major nuclear war, urbanization and industrialization, overpopulation, over-use of water, food, minerals, raw materials and other natural resources.

The build-up of carbon dioxide in the atmosphere is now a well-known fact. It is a world-wide process that probably dates back to the beginnings of the industrial revolution, when coal, oil and natural gas — the fossil fuels — first began to be burned on a large scale. Currently the world consumes such fuels at the rate of five gigatonnes of carbon (5 000 million metric tons) per annum. It is possible that this rate will increase. About three gigatonnes of this carbon remain in the atmosphere as carbon dioxide, being added to the 725 gigatonnes already present.

Other trace gases, notably nitrous oxide, methane and various synthetic substances, are also increasing as the result of human action. Because carbon dioxide and these other gases are strong absorbers of infra-red radiation at wavelengths emitted by the Earth, their action impedes the escape of energy to space, hence they tend to raise surface temperatures. A doubling of carbon dioxide may lead (allowing for various feedbacks) to an increase in world surface air temperature of 1.5 to 4.5°C. In the mainly subtropical arid zone, an increase of 2.3°C is not unlikely, while surface net radiative energy input may increase by perhaps three to four per cent. It has recently been calculated that the other trace gases may actually double the direct carbon dioxide effect. Recent trends in global surface temperature anomalies are illustrated in Figure 6.

Such a warming will be enough to alter greatly the hydrological cycle. Some modelling exercises predict that the changes in heating will considerably increase the rainfall in most parts of the tropics and subtropics. What is still quite uncertain is the potential effect on the great monsoon wind systems of Africa, Asia and Australia. These are the rain-bringers to the greatest extent of arid land, where populations are often dense. If rainfall increases, it is certain that evapotranspiration will increase also. It is quite uncertain at the present time what the effect will be on soil moisture, streamflow and such things as pest incidence and wildfire.

IPCC assessments indicate that:

- Emissions from human activities are significantly increasing the atmospheric concentrations of the greenhouse gases, which have the potential of keeping the world warmer than otherwise;
- Available evidence from observations and modelling studies indicates that doubling of CO_2 concentrations in the atmosphere is unlikely to increase the mean global temperature outside the range 1.5–4.5°C. The corresponding mean global sea-level rise is estimated to be about 10–30 cm. Changes at local and regional levels are however likely to be beyond these ranges;

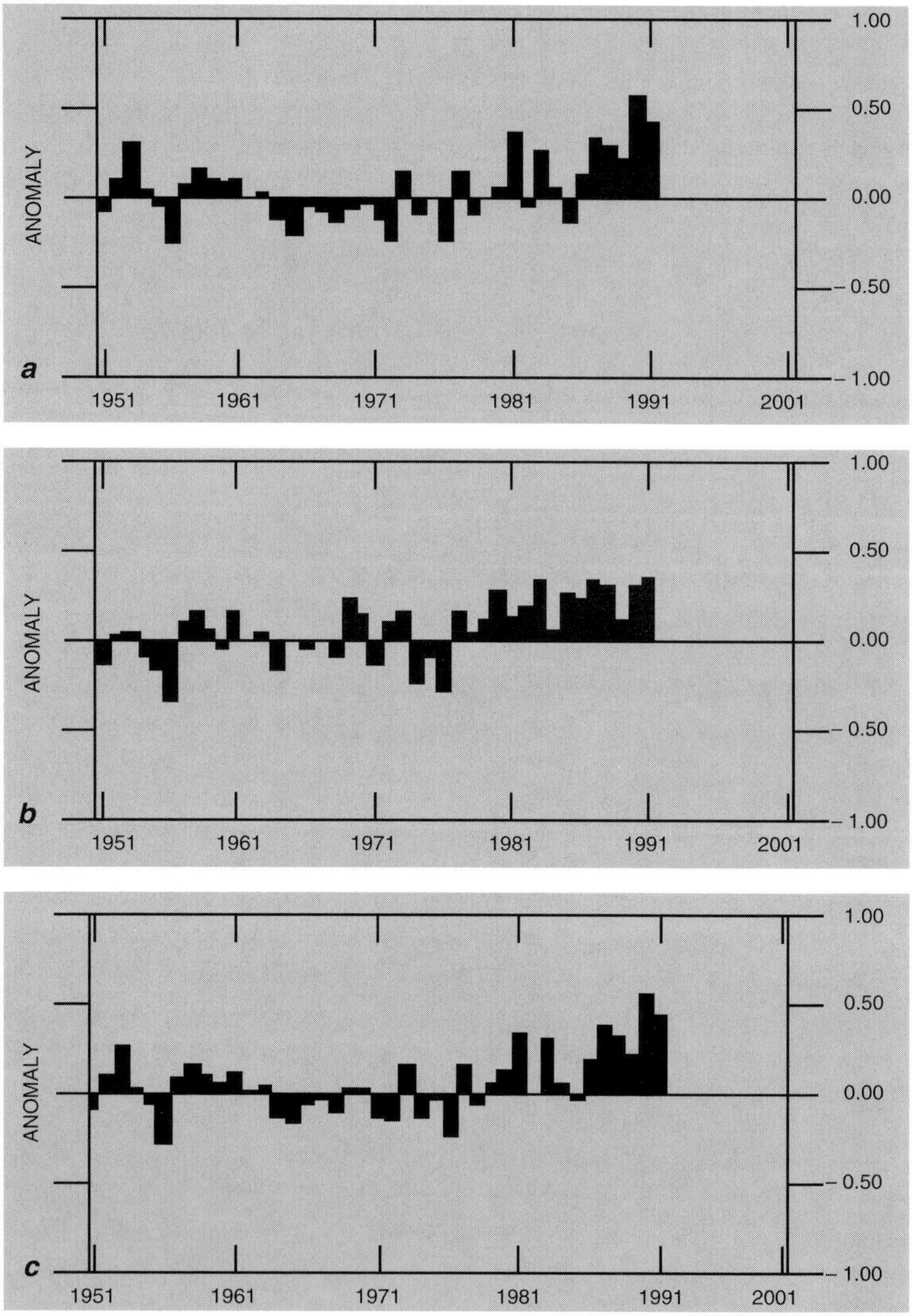

Figure 6 — Time series of estimated (*a*) N.H., (*b*) S.H., and (*c*) global surface temperature anomalies for the calendar year. Temperature anomaly estimates are formed from area-weighted averages of station anomalies over 2° latitude by 2° longitude grids (*After Climate Analysis Center (USA)*)

- There are still many uncertainties in climate predictions, especially at local and regional levels;
- Impacts of any climate changes are severest at local and regional levels. The scale of information of the expected changes is however still very low at these levels;
- It is still difficult to separate natural from human-induced forces in the current global trends of climate;
- The impacts of man on the environment may still take several years to be detected on account of the long period of residence of some of the pollutants, and of the feedback processes;
- The feedback processes between the various climate systems are still not well known.

The impacts of man on the environment are therefore real, and have the potential of changing the natural climate if business continues as usual in terms of the degradation of the natural environment.

Societies living under climates that are marginal and present hazards for human settlement are obviously more at risk than those enjoying a good deal of leeway. Hence it is crucial that the climatologist ascertain as soon as possible whether these foreshadowings are correct. This indicates, indeed, a great need for well-co-ordinated multidisciplinary activities under the World Climate Programme. Some of these are already under way.

It would be wrong to claim, however, that the problem of desertification could be solved if the future course of world climate were fully known. Climates did not create the population pressure now prevalent in so many parts of the arid zone, nor is climatic variability the major cause of the damage that has been done.

As has been said repeatedly above, human economic and political systems — the means of earning a livelihood and of self-government — have been unable to cope with recent climatic fluctuations. Hazards that would easily have been absorbed in earlier times now threaten the livelihood, even the lives, of millions. If desertification is to be brought under control, it is imperative that human societies relearn what they first learned thousands of years ago: that they can prosper in the arid zone only if they can survive its harshest threats and devise an economy in harmony with nature.

It was formerly believed that such economic adaptation was essentially a technical matter. If water was in short supply, the deep groundwater might be tapped; if existing strains of cultivated crops such as millet, sorghum or rice could not withstand very dry conditions, new cultivars more tolerant of drought could be developed. The list of such techniques is very long.

Some of the solutions proposed were in the climatic domain or in the field of hydrology. Thirty years ago this booklet might have dealt almost exclusively with such adaptations of weather, climate and water supply by means of geotechnical measures, e.g. weather modification by means of cloud seeding, large-scale diversions of river systems to provide extended irrigation, flooding of desert basins to increase evaporation and local rainfall, and the planting of greenbelts also to increase rainfall. A proper treatment of these technologies is beyond the scope of this booklet.

Cloud-seeding and rain-making, for example, have turned out to be less practically useful than was once hoped. Although a number of careful analyses have been carried out, it is usually difficult or impossible to demonstrate that useful results have been obtained, often because the necessary measurements are not available. In the arid zone, moreover, drought usually arises from the absence of suitable cloud systems. Cloud-seeding does not create cloud. In limited areas, and under carefully controlled conditions (as has been done in the Negev of Israel), it may be possible to augment local rainfall. As a general measure of relief from drought, however, the technique is not suitable.

Similarly, grandiose schemes for flooding large desert basins are unlikely to increase rainfall significantly. Lake Eyre, in the heart of Australia's dry belt, does occasionally flood itself, and for a while may contribute showers to the surrounding area of semi-desert scrub. Invariably, however, the lake dries out again. Rainfall depends primarily on far-travelled water vapour and on travelling disturbances. It also requires some cooling system for the available water vapour. Flooding depressions in the deserts will do little to alter climate.

It should be noted that, whereas the impact of natural forces on the environment is generally beyond the control of man, that resulting from the mismanagement of human activities is more easily mitigated.

To devise an economy that is in harmony with nature requires more than technological "fixes" of the above kind. This booklet has made the case that land-use control is the key to maintaining healthy and productive microclimates, ecosystems, and food production. Applied climatology and ecology go hand in hand with agricultural science and technology, and with forestry practice, in seeking this harmony.

The future role of the meteorologist in the arid zone must hence be to collaborate closely with scientists working in these domains. Useful results will be achieved only if, at the same time, both social and economic affairs in the countries concerned are in good hands. Political leaders of the affected countries can be sure that meteorologists can and will offer technical guidance to help restore prosperity to the arid zone. Political will is however a major prerequisite for the success of any good technical programme.

The preventive measures for combating drought and halting the spread of the deserts, as highlighted by UNCED in Agenda 21 under Chapter 12, are:

- Strengthening the knowledge base and developing information and monitoring systems for regions prone to desertification and drought, including the economic and social aspects of the fragile ecosystems;

- Combating land degradation through, *inter alia*, intensified soil conservation, afforestation and reforestation activities;
- Developing and strengthening integrated development programmes for the eradication of poverty and promotion of alternative livelihood systems in areas prone to desertification;
- Developing comprehensive anti-desertification programmes and integrating them into national development plans and national environmental planning;
- Developing comprehensive drought-preparedness and drought-relief schemes, including self-help arrangements for drought-prone areas and the design of programmes to cope with environmental refugees;
- Encouraging and promoting popular participation and environmental education, focusing on desertification control and management of the effects of drought.

ANNEX I

Strategy for WMO activities in support of the combat against drought and desertification during the post-UNCED period

(Summary)

At the recent United Nations Conference on Environment and Development (Rio de Janeiro, June 1992), desertification and drought once again received considerable attention, resulting in a request to the UN General Assembly, at its forty-seventh session, to establish an intergovernmental negotiating committee, under the aegis of the General Assembly, for the elaboration of an international convention to combat desertification in those countries experiencing serious drought and/or desertification, particularly in Africa, with a view to finalizing such a convention by June 1994.

Obviously WMO will be required to play an important role in the activities leading to and including the intergovernmental negotiations in connection with the preparation of the convention. As was the case during the negotiations for the Framework Convention on Climate Change, the preparations for this convention would provide another opportunity for WMO and its Members to ensure that their contributions to desertification control and alleviation of the effects of drought are recognized as cost-effective in the planning and implementation of the convention.

More than fifteen years of implementation of the earlier WMO plan of action have shown that the basic measures foreseen by it have been implemented to the benefit of nations in desert- and drought-prone areas.

An updated strategy for WMO activities in support of the combat against desertification and drought has been prepared, taking into account new WMO programmes, the Framework Convention on Climate Change and Chapter 12 of the UNCED Agenda 21. These activities will be undertaken under various WMO programmes such as the World Weather Watch Programme, World Climate Programme, World Climate Research Programme, Tropical Meteorology Research Programme, Atmospheric Research and Environment Programme, Hydrology and Water Resources Programme and Agricultural Meteorology Programme. Participation of other scientific bodies and institutes, such as the WMO/UNEP Intergovernmental Panel on Climate Change (IPCC), the International Council of Scientific Unions (ICSU), the Consultative Group for International Agricultural Research (CGIAR) and regional and national research institutes, will be ensured. A brief summary of the strategy and its activities follows.

1. Monitoring and assessment of drought and desertification

WMO will continue to assist national Meteorological and Hydrological Services to establish, operate and maintain an effective and efficient observation network, strengthen the collection, communication, analysis and dissemination systems with appropriate data management, and provide information adapted to the needs of user communities.

WMO will also assist Members in the evaluation and use of meteorological, climatological and hydrological data and information and in the development of relevant techniques for the assessment of and combat against the effects of drought and desertification.

WMO will continue to support the drought-monitoring centres located in Nairobi for eastern Africa and in Harare for southern Africa, and be responsible for implementing the AGRHYMET programme in the drought-stricken countries in the Sudano-Sahelian zone.

2. Application of meteorological and hydrological data

WMO will continue to assist Members in, and prepare appropriate guidance material for, the application of meteorology and hydrology to technical, scientific and socio-economic development programmes, especially in agriculture and land-use projects, assessment of water resources, and in instituting preparedness, management, response and remedial action against adverse effects of drought and desertification. It will also act as the executing agency for the African Centre for Meteorological Applications for Development, located in Niamey, Niger.

3. Research

Various research programmes are being undertaken for the study of the weather and climate of arid, semi-arid and subhumid and other desert-prone areas with a view to the prediction of long-term trends in the general circulation, various rain-producing atmospheric disturbances, and meteorological drought, using statistical and dynamical methods. These include studies of the *El Niño*-Southern Oscillation phenomenon and other possible causes of meteorological drought and its impact on socio-economic systems.

4. Education and training including public awareness

Training and education will be given top-priority consideration. Through short- and medium-term expert missions, roving and training seminars, training courses and symposia, WMO will continue to provide on-the-job training in operational agrometeorology, and training in specific techniques to assess and minimize the effects of desertification and drought, especially in arid, semi-arid, subhumid and other desert-prone areas.

WMO will also give increased attention to creating public awareness and to communication with a view to bridging the gap between scientific research officers in the areas of desertification and drought and users of meteorological, climatological and hydrological information, thereby improving environmental awareness and increased project impact.

5. Capacity building

WMO will promote the establishment of national agrometeorological services, which should collaborate closely with the user services, often through formal or informal multidisciplinary working groups and institutions, and national climate committees.

6. Conclusion

The future role of WMO will therefore be to assist Members to increase their capabilities in the application of meteorological, climatological and hydrological information in the planning and management of agricultural and other land- and water-use programmes, in order to minimize land degradation and desertification and provide technical and scientific guidance in national development projects. Education, training and capacity building will also receive priority consideration.

ANNEX II

Guidelines for strategy regarding the implementation by WMO Members of UNCED Agenda 21, Chapter 12 on desertification and drought

The UN General Assembly adopted Resolution 32/172 of December 1977 accepting the recommendations of the UN Conference on Desertification (August/September 1977). UNCOD Recommendation 23 in particular was an invitation to WMO to support international action to combat desertification in the context of the UN Plan of Action. This recommendation was considered by the Executive Council at its thirtieth session, which approved a WMO Plan of Action (annex to Resolution 17 (EC-XXX)). After some ten years of implementation of this plan, a new strategy for WMO activities in support of the combat against drought and desertification was adopted at the forty-second session of the Executive Council.

The UN Conference on Environment and Development (June 1992) adopted Agenda 21, Chapter 12 of which deals with drought and desertification. This chapter was a document negotiated by government representatives and is primarily a plan for national implementation with international support. The following guidelines are intended to assist Directors of national Meteorological and Hydrological Services in the implementation of Agenda 21, in particular its Chapter 12. It should be noted that the action proposed in the plan falls within the normal responsibilities of WMO and that work is already being undertaken on many aspects of the plan.

Programme areas

A — Strengthening the knowledge base and developing information and monitoring systems for regions prone to desertification and drought, including the economic and social aspects of these ecosystems

Background

In spite of the WMO World Weather Watch Programme, which has been carried out for many years now, the networks of meteorological and hydrological observing stations continue to be inadequate while systems for collection, analysis and dissemination of the observed data have broken down in many countries. This situation has rendered the capacity of the existing international, regional and national institutions — particularly those in developing countries — to generate and exchange the relevant data, very limited and inadequate. In order to study and understand the dynamics of drought and climate variations and desertification processes, a well-integrated and co-ordinated information and systematic observation system is very important.

Action to be taken

National Meteorological and Hydrological Services should:

(*a*) Promote the establishment, expansion and strengthening of the national and regional networks (over land and the oceans) of meteorological, climatological, agrometeorological and hydrological stations and other monitoring systems to ensure adequate collection of basic information and its communication among national, regional and international centres;

(*b*) Participate in co-ordination committees on drought and desertification that will be established at national and regional levels;

(*c*) Promote the establishment of observing systems for monitoring the effect of agrometeorological and hydrological factors on land degradation and desertification processes;

(*d*) Encourage research work on interaction between climate and land degradation and desertification, including climate prediction;

(*e*) Support the continued implementation of the WMO World Weather Watch Programme and its components, TOGA, etc.;

(*f*) Study and assess the impact of climate and drought on land resources and disseminate the results;

(*g*) Disseminate the climate diagnostic products issued by the World Climate Data Analysis Centres;

(*h*) Support efforts in existing regional and international monitoring systems of FAO and other organizations, such as Earthwatch, the Sahara and Sahel Observatory, the Global Atmospheric Watch, theWMO/ICSU/IOC Global Climatological Observing System, etc., as they relate to land degradation, desertification and drought;

(*i*) Strengthen regional programmes and international co-operation activities such as those of ACMAD, CILSS, IGADD, SADCC and the Arab Maghreb Union.

B — Combating land degradation through, *inter alia*, intensified soil conservation, afforestation and reforestation activities

Background

Assessment of the implementation of the UNCOD Plan of Action to combat desertification has indicated that desertification continues to affect many dryland areas. It is important to take preventive measures in areas which have not yet been affected or are as yet only slightly affected by desertification.

In addition, corrective measures should also be taken to minimize land degradation in rangelands, rainfed cropland and irrigated lands. This can be done effectively if the climate or climatic factors are taken into account in all land-use systems or programmes, in the management of existing natural resources including natural pastures, forests, soils and watershed areas, and in the use of energy sources including alternative sources of energy (solar and wind energy).

Action to be taken

National Meteorological Services in collaboration with agricultural services should:

(*a*) Undertake climate classification/climatic zoning of the countries with a view to providing information relevant to afforestation and soil conservation programmes using drought-resistant fast-growing species, and preventing soil erosion;

(*b*) Promote the use of agrometeorological data for effective water planning and management in existing irrigated cropland in order to avoid salinization and waterlogging, and for assessing primary productivity of natural pastures in order to determine optimum animal-carrying capacity;

(*c*) Undertake measurement of appropriate meteorological elements such as wind and solar radiation in order to assess the potential for use of alternative energy sources and lessen the use of woodfuel;

(*d*) In collaboration and with support from WMO and other relevant UN and bilateral agencies, strengthen the co-ordinating role of regional and subregional intergovernmental organizations such as ACMAD, CILSS, IGADD, SADCC and the Arab Maghreb Union in their fight against dryland degradation;

(*e*) Participate in education and training and public awareness programmes for farmers and pastoralists in the fields of drought and desertification on improved management of land and water resources in drylands.

C — Developing and strengthening integrated development programmes for the eradication of poverty and promotion of alternative livelihood systems in areas prone to desertification

Background

Important factors affecting the traditional livelihood systems (mainly agropastoral systems) in desert- and drought-prone areas are the occurrence of frequent drought and increasing demographic pressure. The two factors usually tend to accelerate the rate of land degradation and desertification. Although there is very little involvement for WMO in this programme area, the Organization and its Member national Meteorological and Hydrological

Services should be considered as members of a team for rural development and their contribution sought in the general national effort to create adequate and sustainable livelihood systems for the population. They can also contribute, through exchange of information on results of basic and applied meteorological and hydrological research, towards alleviating the effects of poverty among rural populations. The contribution of national Meteorological and Hydrological Services will be relevant in enhancing special skills among such communities in areas such as soil and water conservation, water harvesting, agroforestry and small-scale irrigation.

D — Developing comprehensive anti-desertification programmes and integrating them into national development plans and national environmental planning

Meteorology and hydrology provide major support for national development and environmental action plans. This is particularly relevant to arid, semi-arid and subhumid regions where the land resource base is rather fragile. It is therefore very important for national Meteorological and Hydrological Services to strengthen their capabilities to provide the necessary data and information and to initiate a process for implementing and monitoring strategies for the combat against desertification and the alleviation of the effects of drought. Such strategies should be incorporated into national development plans.

It should be mentioned that Agenda 21 has called for the establishment by the UN General Assembly of an Intergovernmental Negotiating Committee (INC) for the elaboration of an international convention to combat desertification in those countries experiencing serious drought and/or desertification, particularly in Africa, with a view to finalizing such a convention by June 1994.

The proposed convention is aimed at ensuring international co-ordination and co-operation in the preparation and implementation of desertification control programmes. It should also include and promote public awareness and education and training programmes through the existing mass media.

Action to be taken

National Meteorological and Hydrological Services should:

(*a*) Endeavour to participate in the negotiation process for the elaboration of the convention, especially those in the affected areas;

(*b*) Develop strategies for the implementation of meteorological and hydrological aspects of the monitoring of desertification, of national development plans and environmental action plans.

In addition, the following guidelines are provided for the attention of WMO and Members in the preparation of the convention on desertification.

E — Developing comprehensive drought-preparedness and drought-relief schemes, including self-help arrangements, for drought-prone areas and designing programmes to cope with environmental refugees

Background

Drought is a recurring phenomenon in most arid, semi-arid and subhumid areas. Drought forecasting is not yet operationally possible, although statistical and dynamical forecasts are being prepared by some WMO Members for certain areas on experimental basis. Therefore, drought-preparedness and drought-relief programmes are essential in order to cope with the disastrous situations that accompany the occurrence of drought. It should be realized that both research into forecasting of drought and early warning systems are far beyond the capability of many Members in drought-prone areas, located mostly in developing regions, and drought and desertification often extend beyond the boundaries of the individual Members concerned.

Action to be taken

National Meteorological and Hydrological Services should:

(*a*) Develop, establish and strengthen meteorological, climatological and hydrological data-collection systems, especially under the World Weather Watch Programme and Hydrological Operational Multipurpose System;

(*b*) Continue to undertake, support or sponsor research into the causes of drought with a view to better understanding drought and desertification, developing methods for monitoring and forecasting drought and thus minimizing the adverse effects of drought. This involves studies into long-term trends in the general circulation, e.g. under the Tropical Meteorology Programme and the ICSU/WMO World Climate Research Programme, and their effects on local and regional rain-producing systems — monsoon circulations, squall-lines, SST and ENSO — and microscale processes, agrometeorological modelling, crop forecasting techniques, etc. under the Agricultural Meteorology Programme;

(*c*) Create public awareness and information on drought and desertification, and strengthen training capabilities to assess drought impact and develop methodologies for drought forecasting;

(*d*) Assist and provide inputs into drought-preparedness and response plans;

(*e*) Arrange for flow of early-warning meteorological and hydrological information to decision-makers and land and water users to enable countries to implement strategies for drought intervention and relief;

(*f*) Promote international co-operation between WMO and other relevant international/regional organizations, e.g. ACMAD, CILSS, FAO, IGADD, IOC, UNEP, UNESCO, UNDRO, WFP, etc.

F — Encouraging and promoting popular participation and environmental education, focusing on desertification control and management of the effects of drought

Background

Experience has shown that the successful implementation of desertification control and drought-response programmes depends to a large extent on popular support, participation and involvement. Therefore, it is essential to include in the syllabus of education and training programmes the subject of public awareness and knowledge concerning desertification and drought. In particular, agriculturists and other land users in rural areas should be targeted in order to take advantage of their indigenous and traditional experience in land, soil and water conservation.

Action to be taken

National Meteorological Services should:

(*a*) Disseminate knowledge, through various WMO training programmes and roving seminars, about applied research results on soil and water issues, appropriate species, agricultural techniques as well as technological know-how;

(*b*) Promote collaboration among various disciplines (meteorology, hydrology and other environmental sciences and development programmes) and promote development of indigenous know-how and technology transfer;

(*c*) Support programmes of international and regional programmes in the fields of desertification and drought, such as those of ACMAD, the Arab Maghreb Union, CILSS, DMCs, IGADD, SADCC including regional and national research institutes, e.g. ACSAD, CAZRI, ICARDA, ICRISAT, the Lanzhou Desert Institute, etc.;

(*d*) Support and strengthen public education in the fields of desertification and drought.

Guidelines for national Meteorological Services in the negotiation process for a desertification convention

1. WMO and its Members should play an active role in the activities leading to the expected intergovernmental negotiations. In so doing, the following four priority areas should be kept in mind:

(*a*) Improvement of monitoring networks and climate observation systems to examine the extent and scope of the problem;

(*b*) Research activities to develop a better understanding of the processes involved and the linkages between these phenomena;

(*c*) Development of potential solutions to the problems; and

(*d*) Increased efforts in education, training, public awareness and wider dissemination of the results.

2. National Meteorological and Hydrological Services (NMHS) should ensure that they are properly represented at the first session of the INC, to be held in Nairobi, in order to provide an authoritative scientific voice and expertise on relevant matters.

3. Members should ensure that climate, meteorology and hydrology are duly included among the topics to be considered by the multidisciplinary panel of experts to be created to assist the *ad hoc* secretariat to be established by the UN General Assembly.

4. Members should contribute appropriately to the work of the INC by designating meteorologists and hydrologists to be included in the delegations to INC sessions.

5. Members may consider establishing National Committees on Desertification with specialists from relevant fields, including climatology and hydrology, as members. Accordingly, Members should designate suitably qualified staff members of NMHS to serve as focal points and on these committees. Wherever possible, these focal points should be members of the national delegations to the INC sessions.

6. The focal points from NMHS and/or any other experts in meteorology and hydrology should emphasize WMO's role and activities in the field of drought and desertification.

7. Emphasis should also be placed on the fact that many activities mentioned in Agenda 21 have already been part and parcel of the normal WMO activities, approved by its governing bodies under various major and sub-programmes of the WMO Scientific and Technical Programmes, even before UNCED, and that these will be continued and strengthened during post-UNCED period. They include the:

- World Weather Watch Programme;
- Tropical Meteorology Programme;
- Agricultural Meteorology Programme;
- Hydrology and Water Resources Programme;
- World Climate Programme;
- Joint WMO/ICSU/IOC Global Climate Observing System;
- Global Atmosphere Watch;
- Intergovernmental Panel on Climate Change.

8. Members, represented by their national delegations to the INC, should ensure that appropriate articles are included in the convention relating to climate, drought and hydrology.

9. Members should give attention to creating public awareness and communication with a view to bridging the gap between scientific research officers in the areas of drought and desertification and users of meteorological, climatological and hydrological information, thereby improving environmental awareness and project impact.

Abbreviations

ACMAD	African Centre of Meteorological Applications for Development
ACSAD	Arab Centre for Studies of Arid Zones and Dry Lands
AGRHYMET	Regional Training Centre for Agrometeorology and Operational Hydrology and their Applications
CAZRI	Central Arid Zone Research Institute, Jodhpur, India
CGIAR	Consultative Group for International Agricultural Research
CILSS	Inter-State Committee on Drought Control in the Sahel
DC/PAC	Desertification Control Programme Activity Centre (UNEP)
DMC	Drought-monitoring Centre
ECA	Economic Commission for Africa
ENSO	*El Niño* /Southern Oscillation
FAO	Food and Agriculture Organization
GCOS	Global Climate Observing System
ICARDA	International Centre for Agricultural Research in Dry Areas
ICRISAT	International Crop Research Institute for the Semi-arid Tropics
ICSU	International Council of Scientific Unions
IGADD	Intergovernmental Authority against Drought and Desertification
INC	Intergovernmental Negotiating Committee
IOC	Intergovernmental Oceanographic Commission (UNESCO)
IPCC	Intergovernmental Panel on Climate Change
NMHS	National Meteorological and Hydrological Services
OCA/PAC	Oceans and Coastal Areas Programme Activity Centre (UNEP)
PET	Potential Evapotranspiration
SADCC	Southern African Development Co-ordination Community
TOGA	Tropical Ocean and Global Atmosphere Programme
UNCED	United Nations Conference on Environment and Development
UNCOD	United Nations Conference on Desertification
UNDRO	Office of the United Nations Disaster Relief Co-ordinator
UNEP	United Nations Environment Programme
WFP	World Food Programme